Sergei Yu. Pilyugin

Introduction to Structurally Stable Systems of Differential Equations

Translated by the Author

Springer Basel AG

Author's address:

Professor Dr. S. Yu. Pilyugin
Department of Mathematics and Mechanics
State University
Petrodvorets, Bibliotechnaya pl. 2
St. Petersburg, 198904 (Russia)

Originally published as:
Vvedenie v grubye sistemy differentsial'nykh uravneniy

Deutsche Bibliothek Cataloging-in-Publication Data

Piljugin, Sergej J.:
Introduction to structurally stable systems of differential equations / Sergei Yu. Pilyugin. Transl. by the author. – Basel; Boston; Berlin: Birkhäuser, 1992
Einheitssacht.: Vvedenie v grubye systemy differencial'nych uravnenij ‹engl.›
ISBN 978-3-0348-9712-9   ISBN 978-3-0348-8643-7 (eBook)
DOI 10.1007/978-3-0348-8643-7

Originally published by Birkhäuser Verlag Basel in 1992
Softcover reprint of the hardcover 1st edition 1992
Printed on acid-free paper
ISBN 978-3-0348-9712-9

Contents

Preface

This book is based on a one year course of lectures on structural stability of differential equations which the author has given for the past several years at the Department of Mathematics and Mechanics at the University of Leningrad.

The theory of structural stability has been developed intensively over the last 25 years. This theory is now a vast domain of mathematics, having close relations to the classical qualitative theory of differential equations, to differential topology, and to the analysis on manifolds. Evidently it is impossible to present a complete and detailed account of all fundamental results of the theory during a one year course.

So the purpose of the course of lectures (and also the purpose of this book) was more modest. The author was going to give an introduction to the language of the theory of structural stability, to formulate its principal results, and to introduce the students (and also the readers of the book) to some of the main methods of this theory.

One can select two principal aspects of modern theory of structural stability (of course there are some conventions attached to this statement). The first one, let us call it the "geometric " aspect, deals mainly with the description of the picture of trajectories of a system; and the second, let us say the "analytic" one, has in its centre the method for solving functional equations to find invariant manifolds, conjugating homeomorphisms, and so forth.

This book, consisting of 13 chapters and an Appendix, is mostly devoted to the "geometric" aspect of the theory, as it has been developed by the traditions of the Leningrad school of differential equations and by the scientific interests of the author.

A functional equation for a conjugating homeomorphism is solved in the book only once—in the proof of the Grobman–Hartman Theorem (to be found in the Appendix). We do not mention methods of symbolic dynamics in the theory of smooth dynamical systems.

This book is an introduction rather than a monograph, which is why the author has tried to give detailed proofs of some of the "folklore" statements that are commonly replaced by the words "it's easy to see ...". For example, extended proofs are given of some details in the Kupka–Smale Theorem. The book contains a proof of the necessity of transversality of stable and unstable manifolds of rest points and closed trajectories for structural stability (the author has never seen a printed version of this proof).

The author has utilized evey means of simplifying the proofs. Consequently, in some places it was assumed that the system under investigation has additional smoothness (as for example, in the case of the λ-Lemma and in the proof of the necessity of transversality). Systems were considered not on manifolds but in domains in $\mathbf{R}^n$ (in the proof of the Kupka–Smale Theorem, in the proof of the structural stability of a hyperbolic set). Some proofs were given under additonal assumptions (as was the case for the Hirsch–Palis–Pugh–Shub Theorem which was proved for systems having no 1-cycles).

The first 3 chapters contain practically no theorems and give an introduction to the language of the theory: the definitions of flows and cascades, the description of equivalence relations, and the introduction of metrics and topologies on spaces of differential equations and on spaces of diffeomorphisms.

The structure of a neighborhood of a hyperbolic rest point of an autonomous system of differential equations is studied in Chapter 4. We prove the Stable Manifold Theorem using Perron's method. The global structure of stable and unstable manifolds of a rest point is described. In Chapter 5 we consider the structure of stable and unstable manifolds for a hyperbolic periodic point of a diffeomorphism and for a hyperbolic closed trajectory of an autonomous system.

In Chapter 6 we study the transversality of maps and manifolds. We describe simple relations between transversality and hyperbolicity, and prove the λ-Lemma with some of its consequences. Chapter 7 gives a detailed sketch of the proof of the Kupka–Smale Theorem. The Closing Lemma of C. Pugh is discussed in Chapter 8 and we give a proof of the C^0-closing Lemma. In Chapter 9 we prove the following statement: a structurally stable system of differential equations is a Kupka–Smale system.

Homoclinic points are studied in Chapter 10. The main result in this chapter is the theorem of the existence of an infinite set of periodic points in an arbitrary neighborhood of a transversal homoclinic point. We prove this theorem in the simplest case, that is in the case of a 2-dimensional diffeomorphism.

Chapter 11 is devoted to Morse–Smale systems. We here analyse the classical Andronov–Pontryagin Theorem and show that the stable and unstable manifolds of nonwandering trajectories of a Morse–Smale system are submanifolds of the phase space.

The main part of the book is Chapter 12 and is devoted to hyperbolic sets. We prove Smale's Spectral Decomposition Theorem. To establish the structural stability of a hyperbolic set we use systematically the analogues of Perron's Theorem on the existence of bounded solutions in a perturbed hyperbolic system and the Stable Manifold Theorem. This approach was developed by Pliss in the monograph [24]. The theorem of the structural stability of a hyperbolic set is applied to prove the Structural Stability Theorem of Anosov and the Ω-Stability Theorem of Smale. Proving the last Theorem we establish directly the upper semi-continuity of the nonwandering set under perturbations instead of traditional use of filtrations.

Chapter 13 is devoted to the analytic strong transversality condition. We study its relation to the geometric strong transversality condition. The results of Sacker–Sell and Mañé are described. We prove that the analytic strong transversality condition implies the hyperbolicity of the nonwandering set. The Appendix contains a proof of the Grobman–Hartman Theorem.

We do not give any special references to the statements included in the basic university courses of mathematics. The list of references is far from being complete. It contains only those books and research papers which are directly mentioned in the text. To study the theory of structural stability more extensively we recommend the books and surveys [14, 18, 22, 24, 35, 39].

The author is deeply grateful to his teacher Professor V.A. Pliss. The cooperation with V. A. Pliss for many years was crucial in the formation of the author's scientific interests and in the appearance of this book.

The English version differs somewhat from the Russian one. Some recent results published after 1988 are included in this English edition.

List of Symbols

$\mathbf{R}^n$- the Euclidean n-space (we write $\mathbf{R}$ for $\mathbf{R}^1$);
$\mathbf{Z}$- the set of integers;
$\mathbf{R}_+$- the set of nonnegative numbers;

$\mathbf{Z}_+ = \mathbf{R}_+ \cap \mathbf{Z}$; For a vector $x = \begin{pmatrix} x_1 \\ \vdots \\ x_n \end{pmatrix} \in \mathbf{R}^n$

$|x| = \sqrt{x_1^2 + \ldots + x_n^2}$- the Euclidean norm;
$< x, y >$- the scalar product of $x, y \in \mathbf{R}^n$;
For a matrix A

$$\|A\| = \max_{|x|=1} |Ax|;$$

E (or E_n)- the unit $n \times n$ matrix;
For a map f of variables $\xi_1, \ldots, \xi_m$ we write $f \in C^{k_1, \ldots, k_m}_{\xi_1, \ldots, \xi_m}$
if f is of class C^{k_j} with respect to ξ_j;
$f \in C^0_\xi$ if f is continuous with respect to ξ;
$f \in C^w$ if f is analytic.
For a map f $\frac{\partial f}{\partial \xi}$- the Jacobi matrix,
Df- the derivative.
For a set X we denote by $\overline{X}$ the closure,
by Int X the interior, by ∂X the boundary.
For $X \subset \mathbf{R}^n$ mes X- the Lebesgue measure;
For a manifold M dim M- the dimension.
$\dot{x} = \frac{dx}{dt}$.
$\square$ denotes the end of the proof.

Chapter 1

Flows and Cascades

1. Consider an autonomous system of differential equations

$$\dot{x} = F(x), \tag{1.1}$$

where $x \in \mathbf{R}^n$. We assume throughout the book that the function $F \in C^r(\mathbf{R}^n)$, $r \geq 1$. Fix an arbitrary point $x_0 \in \mathbf{R}^n$. By the Existence and Uniqueness Theorem there exists a number $h > 0$ such that there is a unique solution $\varphi(t, x_0)$ of system (1.1) defined on $(-h, h)$ and having the following property: $\varphi(0, x_0) = x_0$. The graph of the map $\varphi(t, x_0) : (-h, h) \to \mathbf{R}^n$ is called the integral curve of this solution. The projection of the integral curve on the phase space $\mathbf{R}^n$, i.e. the set $x = \varphi(t, x_0)$, is called the trajectory of system (1.1) with initial conditions $(0, x_0)$. Throughout this book we denote this trajectory by $\varphi(t, x_0)$.

Furthermore, consider the system

$$\dot{x} = G(x), \tag{1.2}$$

where

$$G(x) = \frac{F(x)}{1 + |F(x)|^2}\ .$$

It is evident that $|G(x)| < 1$, so every maximally extended solution of system (1.2) is defined for all $t \in \mathbf{R}$. It is well-known that a parametrized differentiable curve in $\mathbf{R}^n$ is a trajectory of system (1.1) if and only if at each point p the curve is tangent to the vector $F(p)$. For every $x \in \mathbf{R}^n$ the vectors $F(x)$ and $G(x)$ differ by a scalar multiplier. So the trajectories of systems (1.1) and (1.2) are geometrically the same, but they have different parametrizations. The properties of the set of trajectories we are going to study depend slightly on their parametrizations. So we can assume that all trajectories of system (1.1) are defined for all $t \in \mathbb{R}$.

Under our assumptions system (1.1) generates a map $\varphi : \mathbb{R} \times \mathbb{R}^n \to \mathbb{R}^n$ having the following properties:

(i) for every $x \in \mathbb{R}^n$

$$\varphi(0, x) = x, \tag{1.3}$$

(ii) for every $t, s \in \mathbb{R}$, $x \in \mathbb{R}^n$

$$\varphi(t + s, x) = \varphi(t, \varphi(s, x)); \tag{1.4}$$

and

$$\varphi \in C^{r+1,r}_{t,x}. \tag{1.5}$$

Property (1.3) follows from the definition of a trajectory. If we identify $\mathbb{R}^n$ and $0 \times \mathbb{R}^n$ we can write (1.3) as $\varphi|_{0 \times \mathbb{R}^n} = id$. Property (1.4) is the basic property of autonomous systems. Property (1.5) follows from the differentiability of solutions.

Any map φ having the properties (1.3)–(1.5) is called a smooth (of class C^r) flow or a smooth (of class C^r) dynamical system with continuous time on $\mathbb{R}^n$. The set

$$\{x = \varphi(t, x_0) : t \in \mathbb{R}\}$$

is called the trajectory of the point x_0 in the flow φ.

It is well-known that there are the three following possibilities for a trajectory $\varphi(t, x_0)$ of system (1.1):

(1) $\varphi(t, x_0) = x_0$, $t \in \mathbb{R}$; in this case the point x_0 is called a rest point;

(2) the solution $\varphi(t, x_0)$ is periodic having the least positive period; in this case the trajectory is called closed;

(3) for any $t_1, t_2 \in \mathbb{R}$ such that $t_1 \neq t_2$,

$$\varphi(t_1, x_0) \neq \varphi(t_2, x_0).$$

2. Consider a diffeomorphism $f : \mathbb{R}^n \to \mathbb{R}^n$ of class C^r, $r \geq 1$. Define a map $\varphi : \mathbb{Z} \times \mathbb{R}^n \to \mathbb{R}^n$:

for $m > 0$

$$\varphi(m,x) = f^m(x) = \underbrace{f(f(\dots f}_{m \text{ times}}(x)\dots));$$

$$\varphi(0,x) = x;$$

for $m < 0$

$$\varphi(m,x) = f^m(x) = \underbrace{f^{-1}(f^{-1}(\dots f^{-1}}_{|m| \text{ times}}(x)\dots)).$$

It is evident that the map φ has the following properties:

$$\varphi|_{0\times\mathbb{R}^n} = id; \tag{1.6}$$

—for every $m_1, m_2 \in \mathbf{Z}$, $x \in \mathbb{R}^n$

$$\varphi(m_1 + m_2, x) = \varphi(m_1, \varphi(m_2, x)); \tag{1.7}$$

—if we fix $m \in \mathbf{Z}$ then the map

$$\varphi(m,x) : \mathbb{R}^n \to \mathbb{R}^n \text{ is of class } C^r. \tag{1.8}$$

Any map φ having the properties (1.6)–(1.8) is called a smooth (of class C^r) cascade or a smooth (of class C^r) dynamical system with discrete time on $\mathbb{R}^n$. The set

$$\{x = \varphi(k, x_0) : k \in \mathbf{Z}\}$$

is called the trajectory of the point x_0 in the cascade φ (or the orbit of x_0).

It is easy to see that there are two following possibilities for the trajectory $\varphi(k, x_0)$ of a cascade φ :

(1) there exists $m \neq 0$ such that $\varphi(m, x_0) = x_0$; in this case there is the least positive m_0 such that $\varphi(k + m_0, x_0) = \varphi(k, x_0)$ for all $k \in \mathbf{Z}$, the point x_0 is called a periodic point of period m_0, and the trajectory of x_0 consists of m_0 different points (if $m_0 = 1$, then the point x_0 is called a fixed point of φ);

(2) for every $m \neq 0$, $\varphi(m, x_0) \neq x_0$; in this case the trajectory of x_0 is an infinite countable set.

3. Let us describe some connections between systems of differential equations and diffeomorphisms.

3a. Consider a smooth (of class C^r, $r \geq 1$) flow φ generated by system (1.1). Define the following map f: for $x \in \mathbf{R}^n$ let $f(x) = \varphi(1, x)$. We show that the map f is a diffeomorphism. There exists an inverse map: $f^{-1}(x) = \varphi(-1, x)$. Indeed, for any x

$$f^{-1}(f(x)) = \varphi(-1, \varphi(1, x)) = \varphi(0, x) = x.$$

Both the maps f, f^{-1} are of class C^r.

We say in this case that the diffeomorphism f and the cascade generated by this diffeomorphism are embedded in the flow φ. It follows from the definition that the basic properties of trajectories of the diffeomorphism and of the flow are analagous.

Note that there exists diffeomorphisms which cannot be embedded in smooth flows. The set of diffeomorphisms having this property is large enough—this set is residual in the space of all diffeomorphisms [21] (see Chapter 3 for exact definitions and statements).

3b. Fix a point $p \in \mathbf{R}^n$ and suppose that the point p is not a rest point of system (1.1). Fix also a number $t_0 \neq 0$. Let Γ_1, Γ_2 be two $(n-1)$-dimensional smooth surfaces in $\mathbf{R}^n$ such that $p \in \Gamma_1$, $q = \varphi(t_0, p) \in \Gamma_2$; the vectors $F(p), F(q)$ have nonzero angles with surfaces Γ_1, Γ_2 at the points p, q respectively. We say that the surfaces Γ_1, Γ_2 are transversal to the trajectory $\varphi(t, p)$.

Theorem 1.1. *The map T of the surface Γ_1 into the surface Γ_2 generated by the shift along trajectories of system (1.1) is a diffeomorphism of a neighborhood of the point p in Γ_1 onto a neighborhood of the point q in Γ_2.*

To prove Theorem 1.1 we need the result of the Implicit Function Theorem. Consider two Eucidean spaces $\mathbf{R}^{n_1}, \mathbf{R}^{n_2}$ with coordinates x_1, x_2 respectively.

Theorem 1.2. *Let V be a neighborhood of a point $(a, b) \in \mathbf{R}^{n_1} \times \mathbf{R}^{n_2}$ and let $f : V \to \mathbf{R}^{n_2}$ be a C^r map, $r \geq 1$. Suppose that $f(a, b) = 0$*

and that

$$\operatorname{rank}\frac{\partial f}{\partial x_2}(a,b) = n_2.$$

Then there exists a neighborhood U of the point a in $\mathbb{R}^{n_1}$ and a map $g : U \to \mathbb{R}^{n_2}$ of class C^r such that $g(a) = b$ and $f(x, g(x)) = 0$ for $x \in U$.

Proof (of Theorem 1.1). Let the surfaces Γ_1, Γ_2 be parametrized by maps

$$\Phi : \mathbb{R}^{n-1} \to \mathbb{R}^n, \Psi : \mathbb{R}^{n-1} \to \mathbb{R}^n; \quad \Phi, \Psi \in C^r,$$

respectively. Now suppose that Γ_1 is parametrized by parameter $s \in \mathbb{R}^{n-1}$; $\Phi(0) = p$; Γ_2- by parameter $\sigma \in \mathbb{R}^{n-1}$, $\Psi(0) = q$.

The tangent space of Γ_1 at p is spanned by the vectors

$$\begin{pmatrix} \frac{\partial \Phi_1}{\partial s_1}(0) \\ \cdots \\ \frac{\partial \Phi_n}{\partial s_1}(0) \end{pmatrix}, \ldots, \begin{pmatrix} \frac{\partial \Phi_1}{\partial s_{n-1}}(0) \\ \cdots \\ \frac{\partial \Phi_n}{\partial s_{n-1}}(0) \end{pmatrix}, \tag{1.9}$$

here $\Phi_1, \ldots, \Phi_n$ are components of Φ, $s_1, \ldots, s_{n-1}$ are components of s. It follows from our assumption that the vectors $F(p)$,

$$\frac{\partial \Phi}{\partial s_1}(0), \ldots, \frac{\partial \Phi}{\partial s_{n-1}}(0)$$

are lineary independent, so

$$\operatorname{rank}\left(\frac{\partial \Phi}{\partial s}(0), F(p)\right) = n. \tag{1.10}$$

Similarly

$$\operatorname{rank}\left(\frac{\partial \Psi}{\partial \sigma}(0), F(q)\right) = n. \tag{1.11}$$

The trajectory of (1.1) beginning at $\Phi(s)$ intersects Γ_2 if and only if there exist $t \in \mathbb{R}$, $\sigma \in \mathbb{R}^{n-1}$ such that $\varphi(t, \Phi(s)) = \Psi(\sigma)$. Consider the function

$$f(s, t, \sigma) = \varphi(t, \Phi(s)) - \Psi(\sigma).$$

This function is defined in a neighborhood of $(0, t_0, 0) \in \mathbf{R}^{n-1} \times \mathbf{R} \times \mathbf{R}^{n-1}$ and is of class C^r. Evidently

$$f(0, t_0, 0) = \varphi(t_0, p) - q = 0.$$

The Jacobi matrix

$$\begin{aligned}\frac{\partial f}{\partial(t,\sigma)}(0, t_0, 0) &= \left(\frac{\partial f}{\partial t}, \frac{\partial f}{\partial \sigma}\right)(0, t_0, 0)\\ &= \left(\frac{\partial \varphi(t, \Phi(s))}{\partial t}, -\frac{\partial \Psi}{\partial \sigma}\right)(0, t_0, 0)\\ &= \left(F(q), -\frac{\partial \Psi}{\partial \sigma}(0)\right).\end{aligned}$$

It follows from (1.11) that the conditions of Theorem 1.2 are satisfied for f with $\mathbf{R}^{n_1} = \mathbf{R}^{n-1}$, $\mathbf{R}^{n_2} = \mathbf{R} \times \mathbf{R}^{n-1}$, $a = 0$, $b = (t_0, 0)$. Hence there exist maps $t = t(s)$ and $\sigma = \sigma(s)$ of class C^r for small $|s|$ such that $f(s, t(s), \sigma(s)) = 0$, i.e.

$$\varphi(t(s), \Phi(s)) = \Psi(\sigma(s)),$$

and $t(0) = t_0$, $\sigma(0) = 0$. Define the map T in the following way: for $\Phi(s) \in \Gamma_1$ with small $|s|$ let $T(\Phi(s)) = \varphi(t(s), \Phi(s)) = \Psi(\sigma(s)) \in \Gamma_2$. The map T is of class C^r. We use (1.10) to prove similarly that the map T^{-1} is also of class C^r. □

The most important case is the following: the trajectory $\varphi(t, p)$ is closed and corresponds to an ω-periodic solution of (1.1); $t_0 = \omega$; $\Gamma_1 = \Gamma_2 = \Gamma$. In this case the diffeomorphism T is called the Poincaré transformation of the closed trajectory $\varphi(t, p)$ generated by Γ.

3c. Consider a system of differential equations

$$\dot{x} = F(t, x), \tag{1.12}$$

where $x \in \mathbf{R}^n$, $F \in C^{0,r}_{t,x}$, $r \geq 1$. Suppose that there exists $\omega > 0$ such that

$$F(t + \omega, x) \equiv F(t, x).$$

Denote by $x(t, t_0, x_0)$ the solution of system (1.12) with initial conditions (t_0, x_0). Define the following map T. If a solution $x(t, t_0, x_0)$ is defined on $[0, \omega]$, let $T(x_0) = x(\omega, 0, x_0)$. Similarly to §3a, we can show that T is a diffeomorphism on its range of definition. The map T is called the Poincaré transformation of system (1.12). It is easy to see that a point $x_0 \in \mathbb{R}^n$ is a periodic point of period m for T if and only if the solution $x(t, 0, x_0)$ of system (1.12) is periodic and has the least positive period equal to $m\omega$.

4. We shall study flows and cascades not only in $\mathbb{R}^n$ but also on manifolds. Let M be a smooth (of class C^r, $r \geq 1$) n-dimensional manifold. As usual that means the following: M is a Hausdorff topological space with a countable basis having an atlas of class C^r, i.e., a family of pairs (α, W) such that:

(1) in a pair (α, W), W is an open set in M and α is a homeomorphism W onto an open set in $\mathbb{R}^n$ (the pair (α, W) is called a chart);

(2) if $(\alpha_1, W_1), (\alpha_2, W_2)$ are charts with $W_1 \cap W_2 \neq \emptyset$, then the maps

$$\alpha_2 \circ \alpha_1^{-1} : \alpha_1(W_1 \cap W_2) \to \mathbb{R}^n,$$
$$\alpha_1 \circ \alpha_2^{-1} : \alpha_2(W_1 \cap W_2) \to \mathbb{R}^n$$

are diffeomorphisms of class C^r;

(3) for any $x \in M$ there exists a chart (α, W) with $x \in W$.

For a chart (α, W) and $x \in W$ we call $\alpha(x)$ the Euclidean coordinates of x in this chart. We call a manifold M closed if M is compact and has no boundary.

Let M_1, M_2 be manifolds of class C^r. Consider a map $f : M_1 \to M_2$. We call the map f smooth of class $C^s, 1 \leq s \leq r$, at $x \in M_1$ if for any charts (α_1, W_1) in $M_1, (\alpha_2, W_2)$ in M_2 such that $x \in W_1, f(x) \in W_2$ the map $\alpha_2 \circ f \circ \alpha_1^{-1}$ is of class C^s in a neighborhood of $\alpha_1(x)$. We call the map f smooth of class C^s if it is of class C^s at any point $x \in M_1$.

Define now a smooth (of class C^s) curve γ on a manifold M of class C^r, $1 \leq s \leq r$, as a map $\gamma : (a, b) \subset \mathbb{R} \to M$ of class C^s. Consider two curves $\gamma_1 : (a_1, b_1) \to M, \gamma_2 : (a_2, b_2) \to M$ of class C^r and suppose that for $a \in (a_1, b_1)$, $b \in (a_2, b_2)$

$$\gamma_1(a) = \gamma_2(b) = \gamma.$$

We say that γ_1, γ_2 are tangent at γ if for a chart (α, W) with $\gamma \in W$

$$\lim_{t\to a} \frac{\alpha(\gamma_1(t)) - \alpha(\gamma)}{t-a} = \lim_{t\to b} \frac{\alpha(\gamma_2(t)) - \alpha(\gamma)}{t-b} \tag{1.13}$$

(t is the parameter on $\mathbb{R}$). It is easy to see that this definition does not depend on the choice of the chart (α, W). Evidently the relation of tangency is an equivalence relation on the set of all smooth curves on M passing through γ.

We define now a tangent vector of M at $x \in M$ as a class of equivalence of smooth curves passing through x. The tangent space to M at $x \in M$, T_xM is the set of of all tangent vectors of M at x. If the manifold M is n-dimensional, then T_xM has a natural n-dimensional vector space structure (see [3]). Consider a chart (α, W), a point $x \in W$, and a tangent vector $v \in T_xM$. Take a smooth curve $\gamma : (a, b) \to M$ in the class of equivalence v such that $\gamma(t_0) = x$. Consider a vector $u \in \mathbb{R}^n$ defined by

$$u = \lim_{t\to t_0} \frac{\alpha(\gamma(t)) - \alpha(x)}{t - t_0}.$$

We define the map $A_x : T_xM \to \mathbb{R}^n$ by $A_xv = u$. This map is a linear isomorphism between T_xM and $\mathbb{R}^n$.

The set of all pairs $(x, v) : x \in M, v \in T_xM$ is called the tangent bundle TM of the manifold M. TM is a smooth manifold of class C^{r-1} and dim $TM = 2n$. Fix $(x, v) \in TM$ and a chart (α, W) such that $x \in W$. Consider the following set

$$TW = \{(\xi, w) : \xi \in W, w \in T_\xi M\}$$

and define the map $\beta : W \to \mathbb{R}^{2n}$ by

$$\beta(\xi, w) = (\alpha(\xi), A_\xi W) \text{ for } (\xi, w) \in W.$$

It is easy to see that (β, TW) is a chart for TM.

The map $p : TM \to M$ defined by $p(x, v) = x$ is called the projection. It is evidently a smooth map of class C^{r-1}.

A tangent vector field of class C^s on a smooth manifold M of class C^r, $1 \le s \le r - 1$, is a map $F : M \to TM$ such that $p \circ F = id$. The last equality means that $F(x) = (x, v)$, where $v \in T_xM$. In other words, to define a vector field on M we fix for every $x \in M$ a class of smooth curves such that any two curves in this class are tangent at x. We shall denote the vector v in a pair $F(x) = (x, v)$ also by $F(x)$ (this will not lead to a confusion).

A vector field F on M generates a system of differential equations

$$\dot{x} = F(x) \tag{1.14}$$

in the following way. We say that a smooth curve $\varphi : (a, b) \to M$ is a trajectory of the vector field F (and also of system (1.14)) if for any $t \in (a, b)$ the curve $\varphi(t)$ is in the class of equivalence $F(\varphi(t))$.

Consider a vector field F of class C^r, $r \geq 1$. Fix a point $x_0 \in M$ and a chart (α, W) such that $x_0 \in W$. Define a vector-function on $\alpha(W) \subset \mathbb{R}^n$ by

$$G(\xi) = A_{\alpha^{-1}(\xi)}\, F(\alpha^{-1}(\xi)).$$

By the Existence and Uniqueness Theorem, system

$$\dot{\xi} = G(\xi)$$

has a unique solution $g(t, \alpha(x_0))$ with initial conditions $(0, \alpha(x_0))$ on an interval $(-h, h)$ for some $h > 0$.

The curve $\varphi(t, x_0) = \alpha^{-1}(g(t, \alpha(x_0)))$ is a trajectory of system (1.14).

If the manifold M is compact, then every maximally extended trajectory of a vector field F is defined for $t \in \mathbb{R}$ [3]. Thus a smooth vector field (and a corresponding system of differential equations) generates on a compact manifold M a smooth flow φ, i.e., a map $\varphi : \mathbb{R} \times M \to M$ having properties analagous to properties (1.3)–(1.5).

If $f : M \to M$ is a diffeomorphism of a manifold M, it generates a cascade on M in an analagous way to the one described in §2.

Chapter 2

Equivalence Relations

1. Consider two systems of differential equations in $\mathbb{R}^n$:

$$\dot{x} = F(x) \tag{2.1}$$

and

$$\dot{x} = G(x). \tag{2.2}$$

We assume that these systems satisfy all conditions formulated in Chapter 1 for system (1.1). Denote by φ, ψ flows generated by systems (2.1), (2.2) respectively.

We say that systems (2.1), (2.2) (and also flows φ, ψ) are topologically equivalent if there exists a homeomorphism $h : \mathbb{R}^n \to \mathbb{R}^n$ which takes trajectories of (2.1) to trajectories of (2.2) preserving their orientation. That means that there exists a scalar function τ on $\mathbb{R} \times \mathbb{R}^n$ having the following properties:

(1) for any $x \in \mathbb{R}^n$ the function $\tau(t, x)$ increases with respect to t and maps $\mathbb{R}$ onto $\mathbb{R}$;

(2) for any $x \in \mathbb{R}^n$, $\tau(0, x) = 0$;

(3) for any $(t, x) \in \mathbb{R} \times \mathbb{R}^n$

$$h(\varphi(t, x)) = \psi(\tau(t, x), h(x)). \tag{2.3}$$

We say that the homeomorphism h is a topological equivalence between systems (2.1) and (2.2).

It follows from the continuity of φ, ψ and from (2.3) that the function τ is continuous. A homeomorphism maps a point onto a point, a closed curve onto a closed curve. So for a rest point p of system (2.1),

$h(p)$ is a rest point of system (2.2), for a closed trajectory γ of system (2.1) , $h(\gamma)$ is a closed trajectory of system (2.2).

Take $\varepsilon > 0$. We say that the systems (2.1), (2.2) (and also flows φ, ψ) are ε-topologically equivalent if there exists a topological equivalence h such that

$$|x - h(x)| < \varepsilon, \; x \in \mathbf{R}^n. \tag{2.4}$$

We say that systems (2.1), (2.2) (and also flows φ, ψ) are topologically conjugate if there exists a homeomorphism $h : \mathbf{R}^n \to \mathbf{R}^n$ such that for any t, x

$$h(\varphi(t,x)) = \psi(t, h(x)). \tag{2.5}$$

We say in this case that the homeomorphism h is a topological conjugacy between systems (2.1), (2.2).

There is an equivalent form of the definition of topological conjugacy; for any t the diagram

$$\begin{array}{ccc} \mathbf{R}^n & \overset{\varphi(t,x)}{\longrightarrow} & \mathbf{R}^n \\ h\downarrow & & \downarrow h \\ \mathbf{R}^n & \overset{\psi(t,x)}{\longrightarrow} & \mathbf{R}^n \end{array}$$

is commutative.

Clearly topologically conjugate systems are topologically equivalent, and $\tau(t,x) \equiv t$ in this case.

It follows from the definitions that the relations of topological equivalence and of topological conjugacy are equivalent relations; they are reflexive, symmetric, and transitive. The relation of ε-topological equivalence is not transitive. The relation of topological conjugacy is too restrictive for the qualitative theory of flows.

Consider the following example. Consider two systems of differential equations in the plane

$$\dot{x} = -y, \; \dot{y} = x, \tag{2.6}$$

$$\dot{x} = -2y, \; \dot{y} = 2x. \tag{2.7}$$

These systems have rest points in the origin. All other trajectories of the systems are circles centered at the origin. If we take an initial point $(x_0, 0)$ then the trajectories of systems (2.6) and (2.7) are given respectively by

$$\psi(t, x_0, 0) : \begin{matrix} x = x_0 \cos t \\ y = x_0 \sin t \end{matrix}; \quad \varphi(t, x_0, 0) : \begin{matrix} x = x_0 \cos 2t, \\ y = x_0 \sin 2t. \end{matrix}$$

Evidently the systems are topologically equivalent; take $h(x, y) = (x, y)$, and $\tau(t, x, y) = 2t$. But they are not topologically conjugate. To get a contradiction suppose that there exists a homeomorphism $h : \mathbf{R}^2 \to \mathbf{R}^2$ such that (2.5) holds. The trajectory $\varphi(t, 1, 0)$ is closed, so its image is also a closed trajectory, and if $h(1, 0) = (x_0, y_0)$, then

$$(x_0, y_0) \neq (0, 0). \tag{2.8}$$

It follows from the definition of φ that $\varphi(\pi, 1, 0) = (1, 0)$, so we have

$$\begin{aligned} (x_0, y_0) &= h(1, 0) = h(\varphi(\pi, 1, 0)) = \\ &= \psi(\pi, h(1, 0)) = \psi(\pi, x_0, y_0) = (-x_0, -y_0). \end{aligned}$$

We have a contradiction with (2.8).

2. In the case of diffeomorphisms the basic equivalence relation is topological conjugacy.

We say that diffeomorphisms $f, g : \mathbf{R}^n \to \mathbf{R}^n$ are topologically conjugate if there exists a homeomorphism $h : \mathbf{R}^n \to \mathbf{R}^n$ such that

$$h \circ f = g \circ h. \tag{2.9}$$

We can give an equivalent form of (2.9):

$$f = h^{-1} \circ g \circ h. \tag{2.10}$$

If (2.9) holds, then for any $k \in \mathbf{Z}$

$$h \circ f^k = g^k \circ h \tag{2.11}$$

We prove (2.11) for $k > 0$ using induction; for $k < 0$ the proof is the same. Suppose that (2.11) holds for $k \leq m$, then

$$\begin{aligned} f^m &= h^{-1} \circ g^m \circ h, \\ f^{m+1} &= f^m \circ f = h^{-1} \circ g^m \circ h \circ h^{-1} \circ g \circ h = h^{-1} \circ g^{m+1} \circ h. \end{aligned}$$

Take $\varepsilon > 0$. We say that diffeomorphisms $f, g : \mathbf{R}^n \to \mathbf{R}^n$ are ε-topologically conjugate is there exists a topological conjugacy h such that (2.4) holds.

3. We define now nonwandering points. Consider a flow $\varphi : \mathbf{R} \times \mathbf{R}^n \to \mathbf{R}^n$.

We say that a point x_0 is a nonwandering point of φ if for any neighborhood U of x_0 and for any $T > 0$ there exist a point $x \in U$ and a number t such that

$$\varphi(t, x) \in U, \ |t| \geq T.$$

It is easy to see that x_0 is nonwandering if and only if there exist sequences $x_k \in \mathbf{R}^n$, $t_k \in \mathbf{R}$ such that

$$x_k \underset{k\to\infty}{\longrightarrow} x_0, \ |t_k| \underset{k\to\infty}{\longrightarrow} +\infty, \ \varphi(t_k, x_k) \underset{k\to\infty}{\longrightarrow} x_0. \tag{2.12}$$

We write Ω_φ (or simply Ω) for the set of nonwandering points of φ. It follows immediately from the definition that the set Ω_φ is closed and invariant by φ. We call trajectories of the set Ω_φ nonwandering. Often the set Ω_φ is also called nonwandering. Simple examples of nonwandering trajectories are rest points and closed trajectories. These nonwandering trajectories are called trivial.

Let us consider an example of a nontrivial nonwandering trajectory.

Denote by ω_x the ω-limit set of the trajectory $\varphi(t, x)$:

$$\omega_x = \{y = \lim_{k\to\infty} \varphi(t_k, x) : t_k \underset{k\to\infty}{\longrightarrow} +\infty\},$$

and by α_x the α-limit set of $\varphi(t, x)$:

$$\alpha_x = \{y = \lim_{k\to\infty} \varphi(t_k, x) : t_k \underset{k\to\infty}{\longrightarrow} -\infty\}.$$

It is well-known that the sets w_x, α_x are closed and invariant.

Lemma 2.1 *For any x, $\alpha_x \cup \omega_x \subset \Omega$.*

Proof. Consider a point $x_0 \in \omega_x$. There exists a sequence $t_l \underset{l\to\infty}{\longrightarrow} +\infty$ such that

$$y_l = \varphi(\tau_l, x) \underset{l\to\infty}{\longrightarrow} x_0.$$

For each l take $k(l)$ such that $\tau_{k(l)} - \tau_l > l$. Consider points $x_l = \varphi(\tau_{k(l)} - \tau_l, y_l)$ and numbers $t_l = \tau_{k(l)} - \tau_l$. Evidently $x_l = y_{k(l)}$ so,

$$y_l \underset{l\to\infty}{\longrightarrow} x_0, \qquad t_l \underset{l\to\infty}{\longrightarrow} +\infty, \qquad x_l = \varphi(t_l, y_l) \underset{l\to\infty}{\longrightarrow} x_0$$

i.e., (2.12) holds. Similarly $\alpha_x \in \Omega$. □

Consider an autonomous system on $\mathbb{R}^2$ whose orbit structure is shown in Figure 1. In this example p, q are saddle rest points, the rest point 0 is unstable. The set ω_z is the union of the closures of trajectories $\varphi(t, r_1), \varphi(t, r_2)$. It follows from Lemma 2.1 that $r_1 \in \Omega$.

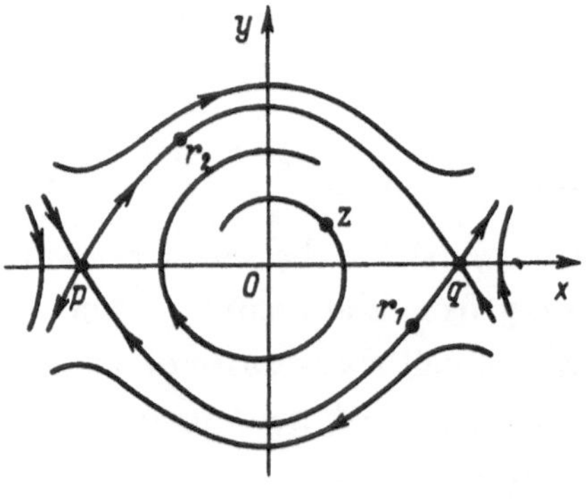

Figure 1.

Note that

$$\varphi(t, r_1) \underset{t\to-\infty}{\longrightarrow} q, \quad \varphi(t, r_1) \underset{t\to+\infty}{\longrightarrow} p.$$

That means that the nonwandering point r_1 while moving along the trajectory $\varphi(t, r_1)$ does not return into a small neighborhood of its initial position. This is not so in the case of trivial nonwandering trajectories.

Let φ, ψ be the flows generated by (2.1), (2.2), and let $\Omega_\varphi, \Omega_\psi$ be the nonwandering sets of these flows. We say that the flows φ, ψ are Ω-equivalent if there exists a homeomorphism h which maps Ω_φ onto Ω_ψ and takes trajectories of (2.1) to trajectories of (2.2) preserving their orientation. We call the homeomorphism h, Ω-equivalence.

Take $\varepsilon > 0$. We say that the flows φ, ψ are ε-Ω-equivalent if there exists an Ω-equivalence h such that (2.4) holds for $x \in \Omega_\varphi$.

Lemma 2.2 *Let h be a topological equivalence of flows φ, ψ, and let $x_0 \in \Omega_\varphi$. Then $h(x_0) \in \Omega_\psi$.*

Proof. Consider sequences x_k and t_k such that (2.12) holds. Passing to a subsequence, we can choose t_k such that $t_k \to +\infty$ or $t_k \to -\infty$ as $k \to \infty$. Suppose that $t_k \to +\infty$. It follows from the continuity of h that

$$h(x_k) \underset{k\to\infty}{\longrightarrow} h(x_0),$$

$$h(\varphi(t_k, x_k)) = \psi(\tau(t_k, x_k), h(x_k)) \underset{k\to\infty}{\longrightarrow} h(x_0).$$

It is sufficient now to prove that $\tau(t_k, x_k) \underset{k\to\infty}{\longrightarrow} +\infty$. To get a contradiction suppose that there exists a bounded subsequence of $\tau(t_k, x_k)$. For simplicity we suppose that $\tau(t_k, x_k) \le \textcircled{H}$. It follows from property (1) of τ that $\tau(t, x_0) \to +\infty$ as $t \to +\infty$. Find $T > 0$ such that $\tau(T, x_0) = \textcircled{H} + 2$. As τ is continuous, $\tau(t, x_k) \underset{k\to\infty}{\longrightarrow} \tau(T, x_0)$, so $\tau(T, x_k) \ge \textcircled{H} + 1$ for large k. The function τ increases with respect to t, $t_k \to +\infty$ as $k \to \infty$, so for large k, $t_k \ge T$, and $\tau(t_k, x_k) \ge \textcircled{H} + 1$. □

Corollary. *Topological equivalence of flows implies their Ω-equivalence, ε-topological equivalence implies ε-Ω-equivalence.*

Analagous concepts are introduced for diffeomorphisms and cascades.

4. We can define topological equivalence and other relations described earlier not on the whole phase space but on its subsets; these subsets need not be invariant.

Let system (2.1) be defined on a domain $G_1 \subset \mathbb{R}^n$, and let system (2.2) be defined on a domain $G_2 \subset \mathbb{R}^n$. We say that system (2.1) on G_1 and system (2.2) on G_2 are topologically equivalent if there exists a homeomorphism h mapping G_1 onto G_2 such that h takes intersections of trajectories of (2.1) with G_1 to intersections of trajectories of (2.2) with G_2 and preserves orientation on trajectories. If $G_1 = G_2 = G$ we say that the systems are topologically equivalent on G.

Definitions of topological conjugacy on G etc. are analagous.

If I_1, I_2 are invariant sets for systems (2.1), (2.2), respectively, we say that the set I_1 is locally topologically equivalent to I_2 if systems

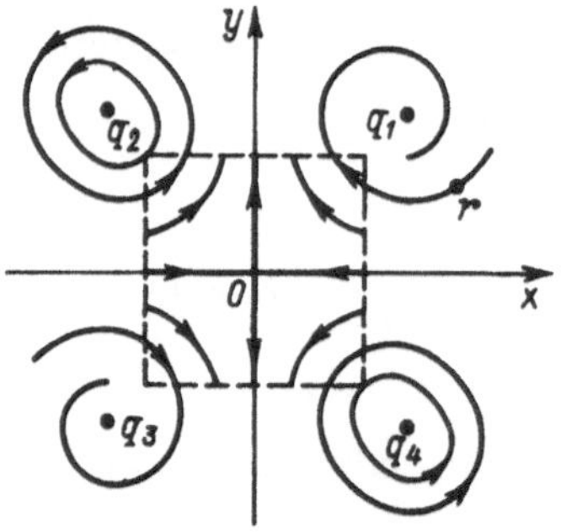

Figure 2

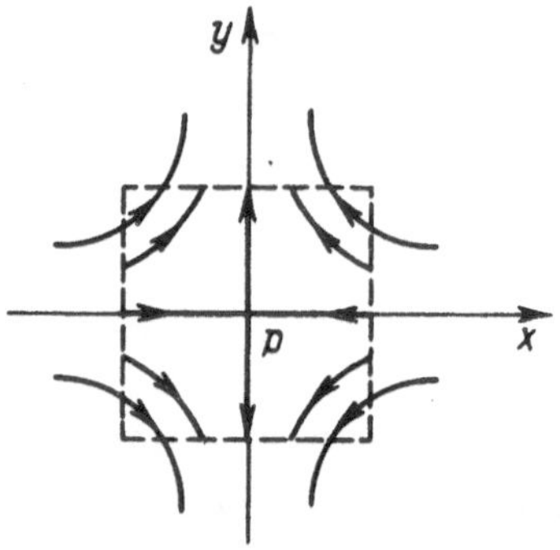

Figure 3

(2.1), (2.2) are topologically equivalent on some neighborhoods U_1, U_2 of I_1, I_2.

Consider for example two autonomous systems on $\mathbb{R}^2$ whose orbit structures are shown in Figures 2,3.

The system whose orbit structure is shown in Figure 2, has 5 rest points: 0 is a saddle, q_2 and q_4 are centers, q_1 and q_3 are asymptotically stable rest points. The system whose orbit structure is shown in Figure 3, has a unique saddle rest point p. (For example take the system

$$\dot{x} = -x, \dot{y} = y.)$$

The saddle point 0 of the first system is locally topologically equivalent to the saddle point p of the second system; the corresponding neighborhoods U_1, U_2 are dotted squares. Note that the intersection of the trajectory of the point r in Figure 2 with U_1 consists of a countable set of components. If h is a local topological equivalence, the h-images of these components belong to different trajectories of the second system.

5. For flows and cascades on a manifold M the definitions of equivalence relations (topological equivalence etc.) are analogous.

Chapter 3

Spaces of Systems of Differential Equations and Diffeomorphisms

1. Let G be a domain in $\mathbb{R}^n$ such that:

(1) its closure $\overline{G}$ is compact;

(2) the boundary S of G is a smooth $(n-1)$-dimensional manifold in $\mathbb{R}^n$ (possibly not connected).

Consider a system of differential equations

$$\dot{x} = F(x) \tag{3.1}$$

where $F \in C^1(\mathbb{R}^n)$. We identify the system (3.1) and its vector field F. Consider also a system

$$\dot{x} = F_1(x). \tag{3.2}$$

Introduce the following equivalence relation: $F \sim F_1$, if $F(x) = F_1(x)$ for $x \in G$. Define the space $X(G)$: elements of $X(G)$ are corresponding classes of equivalence. For simplicity we speak about systems in $X(G)$ regarding them as representatives of classes of equivalence. For two systems $F_1, F_2 \in X(G)$ we define

$$\rho_0(F_1, F_2) = \sup_{x \in G} |F_1(x) - F_2(x)|,$$

$$\rho_1(F_1, F_2) = \rho_0(F_1, F_2) + \sup_{x \in G} \|\frac{\partial F_1}{\partial x}(x) - \frac{\partial F_2}{\partial x}(x)\|.$$

It is easy to see that ρ_0, ρ_1 are metrics on $X(G)$. It is evident that $\rho_0, \rho_1 \geq 0$. If $\rho_0(F, F_1) = 0$ or $\rho_1(F, F_1) = 0$ then $F(x) = F_1(x)$

for $x \in G$, i.e., systems F, F_1 are in the same class of equivalence. Symmetry and the triangle inequality are evident.

Denote by $X^0(G)$ the space $X(G)$ with the metric ρ_0, and by $X^1(G)$ the space $X(G)$ with metric ρ_1. We denote by $X^0(G), X^1(G)$ also the corresponding topological spaces. It is easy to see that the metric space $X^1(G)$ is complete.

Now consider a subset $X_+(G)$ of $X(G)$ consisting of systems (3.1) having the following properties:

(1) $F(x) \notin T_x S, x \in S$;

(2) $\varphi(t, x) \in G$ for $x \in S$ and small $t > 0$.

The properties (1) and (2) mean that for $x \in S$ the vector $F(x)$ is directed into G. It is easy to see that for $F \in X_+(G)$ and $x \in G$, $t \geq 0$ we have $\varphi(t, x) \in G$. So the system $F \in X_+(G)$ generates a map $\varphi : \mathbb{R}_+ \times G \to G$ having the properties (1.3)–(1.5). We call such a map a semiflow on G.

As $\overline{G}$ is compact, the manifold S is also compact. For a system $F \in X_+(G)$ the angle between $F(x)$ and $T_x S$, $x \in S$, is continuous with respect to x, so this angle is separated from zero. It follows immediately that if $\rho_0(F, F_1)$ is small enough, then $F_1 \in X_+(G)$. So $X_+(G)$ is open in $X^0(G)$ (and also in $X^1(G)$). Denote

$$X_+^i(G) = X^i(G) \cap X_+(G), \quad i = 0, 1.$$

2. We define now analogous concepts for diffeomorphisms. Consider a domain $G \subset \mathbb{R}^n$ such that $\overline{G}$ is compact. For two diffeomorphisms $f, f_1 : \mathbb{R}^n \to \mathbb{R}^n$ we write $f \sim f_1$ if $f(x) = f_1(x)$ for $x \in G$. Define the space Diff(G): elements of this space are corresponding classes of equivalence. We speak again about diffeomorphisms in Diff(G) (instead of their classes of equivalence). For $f, f_1 \in$ Diff(G) define

$$\rho_0(f, f_1) = \sup_{x \in G} |f(x) - f_1(x)|,$$

$$\rho_1(f, f_1) = \rho_0(f, f_1) + \sup_{x \in G} \|\frac{\partial f}{\partial x}(x) - \frac{\partial f_1}{\partial x}(x)\|.$$

It is easy to see that ρ_0, ρ_1 are metrics on Diff(G). Denote by Diff$^0(G)$, Diff$^1(G)$ the corresponding metric (and topological) spaces. The space Diff$^1(G)$ is complete.

Consider a subset $\mathrm{Diff}_+(G)$ of Diff (G): $f \in \mathrm{Diff}_+(G)$ if $f(\overline{G}) \subset G$. For $f \in \mathrm{Diff}_+(G)$, $x \in G$ and for $k \geq 0$ we have $f^k(x) \in G$. That means that a diffeomorphism $f \in \mathrm{Diff}_+(G)$ generates a map $\varphi : \mathbf{Z}_+ \times G \to G$ having the properties (1.6)–(1.8). We call such a map a semicascade on G. As G is compact $\mathrm{Diff}_+(G)$ is open in $\mathrm{Diff}^0(G)$ (and in $\mathrm{Diff}^1(G)$). Denote

$$\mathrm{Diff}_+^i(G) = \mathrm{Diff}^i(G) \cap \mathrm{Diff}_+(G), \quad i = 0, 1.$$

3. Let M be a smooth (of class C^r, $r \geq 2$) closed manifold. Denote by $X(M)$ the space of systems of differential equations on M that are generated by vector fields of class C^1 on M.

As M is compact, there exists a finite atlas $(\alpha_1, W_1), \ldots, (\alpha_m, W_m)$. Fix a chart (α_i, W_i), and for $x \in W_i$ consider the map $A_x : T_x M \to \mathbf{R}^n$ defined in Chapter 1, §4. Denote this map by A_x^i. Let F be a vector field of class C^1 on M. Define a vector field F^i on $\alpha_i(W_i)$ by

$$F^i(y) = A_x^i F(x) \qquad \text{for}\, y = \alpha_i(x).$$

Find compact subsets $W_1', \ldots, W_m'$ in domains $W_1, \ldots, W_m$ so that

$$M = W_1' \cup \cdots \cup W_m'.$$

For two vector fields F_1, F_2 define

$$\rho_0(F_1, F_2) = \max_{1 \leq i \leq m} \sup_{y \in \alpha_i(W_i')} |F_1^i(y) - F_2^i(y)|,$$

$$\rho_1(F_1, F_2) = \rho_0(F_1, F_2) + \max_{1 \leq i \leq m} \sup_{y \in \alpha_i(W_i')} \left\| \frac{\partial F_1^i}{\partial y}(y) - \frac{\partial F_2^i}{\partial y}(y) \right\|.$$

It is easy to see that ρ_0, ρ_1 are metrics on $X(M)$. Taking other compact subsets $W_1'', \ldots, W_m''$ or another finite atlas $(\tilde{\alpha}_1, \tilde{W}_1), \ldots, (\tilde{\alpha}_q, \widetilde{W}_q)$ we obtain metrics $\tilde{\rho}_0, \tilde{\rho}_1$. As atlases are finite, the metrics $\rho_0, \tilde{\rho}_0$ induce the same topology on $X(M)$. We denote the corresponding topological space by $X^0(M)$. The space $X^1(M)$ is defined analogously (using the metric ρ_1).

There is an analogous way of introducing topologies in the space of diffeomorphisms of M. We describe another construction.

There is an embedding of the manifold M in $\mathbf{R}^m$ where m is large enough [32]. Consider M as a submanifold in $\mathbf{R}^m$. Denote by Diff(M) the set of diffeomorphisms $f : M \to M$ of class C^1. For $f_1, f_2 \in$ Diff(M) let

$$\rho_0(f_1, f_2) = \sup_{x \in M} |f_1(x) - f_2(x)|,$$

here $| \, |$ is the Euclidean norm in $\mathbf{R}^m$. For $x \in M$, $Df(x)$ is a linear map $T_x M \to T_{f(x)} M$. Let

$$\rho_1(f_1, f_2) = \rho_0(f_1, f_2) + \sup_{x \in M} \|Df_1(x) - Df_2(x)\|,$$

here $\| \, \|$ is the operator norm induced by $| \, |$. We denote by $\text{Diff}^0(M)$, $\text{Diff}^1(M)$ the metric spaces $(\text{Diff}(M), \rho_0)$, $(\text{Diff}(M), \rho_1)$, and also the corresponding topological spaces.

4. We define now structural stability of system (3.1). We consider one of the following cases: $F \in X_+(G)$, where $G \subset \mathbf{R}^n$ (see §1) or $F \in X(M)$, where M is a smooth closed manifold. Let $X = X^1(G)$, $H = \overline{G}$ in the first case, and let $X = X^1(M)$, $H = M$ in the second case.

We say that the system (3.1) is structurally stable if for any $\varepsilon > 0$ there is a neighborhood $U(\varepsilon)$ of the field F in X such that, for any $F_1 \in U(\varepsilon)$, system (3.2) is ε-topologically equivalent to (3.1) on H. This definition corresponds to the original definition given by Andronov and Pontryagin [1].

To define structural stability of a diffeomorphism f we use the concept of toplogical conjugacy. Consider one of the following cases: $f \in \text{Diff}^1_+(G)$, where $G \subset \mathbf{R}^n$, or $f \in \text{Diff}^1(M)$, where M is a smooth closed manifold. Let $X = \text{Diff}^1_+(G)$, $H = \overline{G}$ in the first case and let $X = \text{Diff}^1(M)$, $H = M$ in the second case.

We say that a diffeomorphism f is structurally stable if for any $\varepsilon > 0$ there is a neighborhood $U(\varepsilon)$ of f in X such that any diffeomorphism $f_1 \in U(\varepsilon)$ is ε-topologically conjugate to f on H.

Another important stability concept is connected with the relation of Ω-equivalence. We say that system (3.1) is Ω-stable if for any $\varepsilon > 0$ there is a neighborhood $U(\varepsilon)$ of the field F in X such that, for any $F_1 \in U(\varepsilon)$, system (3.2) is ε-Ω-equivalent to (3.1). An analogous definition is given in the case of a diffeomorphism f. It follows from the Corollary of Lemma 2.2 that structural stability implies Ω-stability.

5. Let X be a metric space with metric ρ. We denote by X also the corresponding topological space. We say that a subset $A \subset X$ is residual if it contains a countable intersection of open dense sets in X. We say that a property of elements of X is generic if it is satisfied by all elements of a residual set in X.

A classical theorem of Baire says that if X is a complete metric space then every residual subset of X is dense in X.

Note that there are residual sets of "small" measure. Let us show that there exists a residual set $V \subset \mathbf{R}$ such that mes $V = 0$. Fix a countable dense set a_n, $n = 0, 1, \ldots$, in $\mathbf{R}$ (for example, take the set of rational numbers). For m natural define the set

$$V_m = \bigcup_{n\geq 0}(a_n - \frac{1}{m \cdot 2^n}, a_n + \frac{1}{m \cdot 2^n}).$$

It is evident that V_m is open and dense in $\mathbf{R}$. As

$$\text{mes } V_m \leq \sum_{n=0}^{\infty} \frac{2}{m \cdot 2^n} = \frac{4}{m}$$

for a residual set $V = \bigcap_{m>0} V_m$ we have mes $V = 0$.

We discussed the embedding of diffeomorphisms in smooth flows in Chapter 1, §3a and can now give the exact statement.

Theorem 3.1. [21] *Let M be a smooth closed manifold, dim $M \geq 2$. A generic diffeomorphism $f \in \mathrm{Diff}^1(M)$ cannot be embedded in a flow generated by a system from $X^1(M)$.*

6. We shall now define some classes of maps-immersions and embeddings [32]. We use in this book immersions of smooth manifolds and embeddings of closed balls in Euclidean spaces.

Let f be a map of a manifold M into a manifold N. We say that f is a C^0 immersion if f is continuous and $x \neq y$ implies $f(x) \neq f(y)$. We say that f is an immersion of class C^k, $k \geq 1$, if f is of class C^k, $x \neq y$ implies $f(x) \neq f(y)$, and for any $x \in M$ rank $Df(x) = \dim M$.

Consider the closed ball $D = \{|x| \leq a\} \subset \mathbf{R}^m$. The ball D is a manifold with boundary ∂D. We say that $f : D \to \mathbf{R}^n, n \geq m$, is an embedding of class C^k if $f(D)$ is a submanifold with boundary in $\mathbf{R}^n$

and $f : D \to f(D)$ is a diffeomorphism of class C^k [32]. We consider in this book manifolds with boundary being images of smooth (of class C^k, $k \geq 1$) embeddings of closed balls, and we call such manifolds with boundary closed smooth balls or closed smooth disks.

Let G be a closed smooth disk. For two embeddings $f_1, f_2 : G \to \mathbf{R}^n$ define

$$\rho_1(f_1, f_2) = \sup_{x \in G} |f_1(x) - f_2(x)| + \sup_{x \in G} \left\| \frac{\partial f_1}{\partial x}(x) - \frac{\partial f_2}{\partial x}(x) \right\|.$$

It is easy to see that ρ_1 is a metric on the space of embeddings $G \to \mathbf{R}^n$. We denote this metric space (and the corresponding topological space) by $E^1(G, \mathbf{R}^n)$. For $G \subset \mathbf{R}^n$ we denote by id the identical embedding $G \to \mathbf{R}^n$. In a similar way we define the space $E^1(G, M)$ of embeddings of G in a manifold M.

Chapter 4

Hyperbolic Rest Point

1. Consider the autonomous system of differential equations (1.1). Let p be a rest point; it is well-known that in this case $F(p) = 0$.

We say that the rest point p is hyperbolic if the eigenvalues λ_j of the matrix

$$A = \frac{\partial F}{\partial x}(p)$$

satisfy the following condition:

$$Re\,\lambda_j \neq 0. \tag{4.1}$$

Fix a hyperbolic rest point p of system (1.1). Suppose that for eigenvalues λ_j of A we have

$$Re\,\lambda_j < 0, j = 1, \ldots, n_1;\ Re\,\lambda_j > 0, j = n_1 + 1, \ldots, n.$$

Denote $n_2 = n - n_1$. If $n_1 = n$ then it is well-known from basic courses in differential equations that the solution $x(t) \equiv p$ is asymptotically stable. That means the following: for each neighborhood U of p there exists a subneighborhood V such that $x \in V$ implies $\varphi(t, x) \in U$ for $t \geq 0$ and $\varphi(t, x) \underset{t \to +\infty}{\longrightarrow} p$ (see Figure 4). In this case we say that the rest point p is attractive. If $n_2 = n$ then the rest point p is attractive for the flow $\varphi(-t, x)$ (See Figure 5).

Let us pay more attention to the case $n_1 n_2 \neq 0$. In this case we say that p is a hyperbolic saddle rest point.

For simplicity we suppose that $p = 0$. We assume that $\mathbb{R}^n$ is decomposed: $\mathbb{R}^n = \mathbb{R}^{n_1} \times \mathbb{R}^{n_2}$ with coordinates y in $\mathbb{R}^{n_1}$, z in $\mathbb{R}^{n_2}$, so that according to this decomposition the matrix A is block-diagonal,

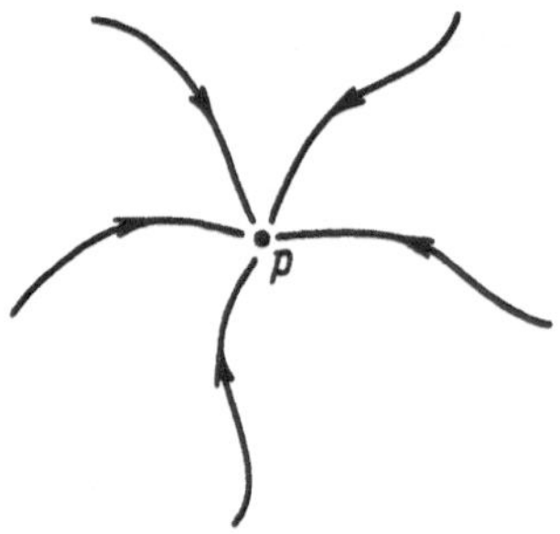

Figure 4.

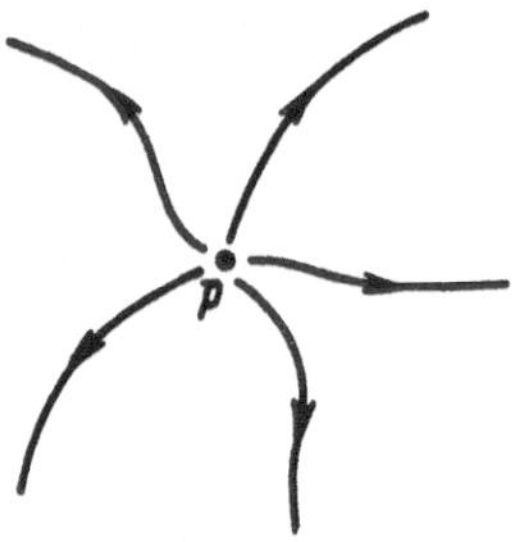

Figure 5.

$$A = \begin{pmatrix} A_1 & 0 \\ 0 & A_2 \end{pmatrix}.$$

Here A_1 is $n_1 \times n_1$ matrix with eigenvalues $\lambda_1, \ldots, \lambda_{n_1}$, A_2 is $n_2 \times n_2$ matrix with eigenvalues $\lambda_{n_1+1}, \ldots, \lambda_n$. This can be achieved by means of linear nonsingular change of variables. In a neighborhood U of the origin, system (1.1) reduces to

$$\begin{aligned} \dot{y} &= A_1 y + F_1(y, z), \\ \dot{z} &= A_2 z + F_2(y, z). \end{aligned} \tag{4.2}$$

The functions F_1, F_2 are of the same class C^r as F in (1.1),

$$F_i(0,0) = 0, \quad \frac{\partial F_i}{\partial(y,z)}(0,0) = 0, \ i = 1, 2. \tag{4.3}$$

We are going to prove now a basic result in the theory of structural stability—the so-called Stable Manifold Theorem (Theorem 4.1 below). Various variants of the theorem were proved by Lyapunov, Hadamard, Perron (see [2] for the history of this theorem). We use Perron's method here. One can use the same proof to construct stable and unstable manifolds of a trajectory in a hyperbolic set (see Theorem 12.1 in Chapter 12).

Theorem 4.1. *Let F_1, F_2 in (4.2) be of class C^1. There exist $\Delta > 0$ and maps*

$$\alpha : \{|y| < \Delta\} \to \mathbb{R}^{n_2},$$
$$\beta : \{|z| < \Delta\} \to \mathbb{R}^{n_1}$$

of class C^1 such that:

(1) $$\alpha(0) = 0,\ \frac{\partial\alpha}{\partial y}(0) = 0; \tag{4.4}$$

(2) $$\beta(0) = 0,\ \frac{\partial\beta}{\partial z}(0) = 0; \tag{4.5}$$

(3) *if x belongs to the set*

$$W^s_{loc}(0) = \{(y,z) : |y| < \Delta, z = \alpha(y)\}$$

then there exists $t_0 \geq 0$ such that

$$\varphi(t,x) \in W^s_{loc}(0) \text{ for } t \geq t_0 \text{ and } \varphi(t,x) \underset{t\to+\infty}{\longrightarrow} 0;$$

(4) *if x belongs to the set*

$$W^u_{loc}(0) = \{(y,z) : y = \beta(z),\ |z| < \Delta\}$$

then there exists $t_0 \leq 0$ such that

$$\varphi(t,x) \in W^u_{loc}(0) \text{ for } t \leq t_0 \text{ and } \varphi(t,x) \underset{t\to-\infty}{\longrightarrow} 0.$$

We shall divide the proof of this theorem into several lemmas.

We denote by $x(t,x_0)$ the solution of system (4.2) with initial conditions $(0,x_0)$. With respect to the decomposition $x = (y,z)$ we set

$$x(t,x_0) = \begin{pmatrix} y(t,y_0,z_0) \\ z(t,y_0,z_0) \end{pmatrix},\ x_0 = \begin{pmatrix} y_0 \\ z_0 \end{pmatrix}.$$

As follows from the basic course of differential equations our assumptions about eigenvalues of matrices A_1, A_2 imply the existence of $a, \lambda > 0$ such that

$$\|e^{A_1 t}\| \leq a e^{-\lambda t}, t \geq 0; \tag{4.6}$$

$$\|e^{A_2 t}\| \leq a e^{\lambda t}, t \leq 0. \tag{4.7}$$

Take l_0 such that the inequality

$$0 < l_0 < \frac{3\lambda}{16a} \tag{4.8}$$

holds. Choose $\varepsilon > 0$ so small that the set

$$U_\varepsilon = \{(y,z) : |y| \le \varepsilon, |z| \le \varepsilon\}$$

is a subset of U and inequalities

$$\left\|\frac{\partial F_1}{\partial x}\right\| < l_0, \left\|\frac{\partial F_2}{\partial x}\right\| < l_0$$

hold for $x \in U_\varepsilon$. It follows from (4.3) that such ε exists. Denote by l the Lipschitz constant of F_1, F_2 in U_ε (evidently, $l \le l_0$), and let $\sigma = 0,5\lambda$. Take $\Delta > 0$ such that

$$2a\Delta < \varepsilon. \tag{4.9}$$

We begin constructing a map α.

Lemma 4.1. *There exists a continuous map $\alpha : \{|y| < \Delta\} \to \mathbb{R}^{n_2}$ having the following property: for each $y_0, |y_0| < \Delta$, there exists $z_0 = \alpha(y_0) \in \mathbb{R}^{n_2}$ such that*

$$|y(t, y_0, z_0)| \le 2a|y_0|e^{-\sigma t}, t \ge 0; \tag{4.10}$$

$$|z(t, y_0, z_0)| \le 2a|y_0|e^{-\sigma t}, t \ge 0. \tag{4.11}$$

Proof. Take a vector $\eta \in \mathbb{R}^{n_1}$, $|\eta| < \Delta$. Suppose that $y(t), z(t)$ is a solution of the system of integral equations

$$\begin{aligned} y(t) &= e^{A_1 t}\eta + \int_0^t e^{A_1(t-s)} F_1(y(s), z(s))ds, \\ z(t) &= -\int_t^\infty e^{A_2(t-s)} F_2(y(s), z(s))ds \end{aligned} \tag{4.12}$$

such that $(y(t), z(t)) \in U_\varepsilon$. Differentiating (4.12) we see that in this case $y(t), z(t)$ is a solution of (4.2). We are going to prove the existence of a solution of (4.12) using a method of successive approximations. Let

$$x_0(t) = \begin{pmatrix} y_0(t) \\ z_0(t) \end{pmatrix} \equiv \begin{pmatrix} 0 \\ 0 \end{pmatrix}, \tag{4.13}$$

and for $m \geq 1$ let

$$x_m(t) = \begin{pmatrix} y_m(t) \\ z_m(t) \end{pmatrix}$$

where

$$y_m(t) = e^{A_1 t}\eta + \int_0^t e^{A_1(t-s)} F_1(y_{m-1}(s), z_{m-1}(s))ds,$$

$$z_m(t) = -\int_t^\infty e^{A_2(t-s)} F_2(y_{m-1}(s), z_{m-1}(s))ds. \tag{4.14}$$

We claim that all the approximations $x_m(t)$ exist and that

$$|x_m(t)| \leq 2a|\eta|e^{-\sigma t}, t \geq 0. \tag{4.15}$$

We prove (4.15) by induction. For $m = 0$ (4.15) evidently holds. Suppose that (4.15) holds for x_{m-1}. Then

$$\begin{aligned} |y_m(t)| &\leq a|\eta|e^{-\lambda t} + \int_0^t 2a^2|\eta|le^{-\lambda(t-s)}e^{-\sigma s}ds \\ &\leq a|\eta|e^{-\lambda t} + 2a^2|\eta|le^{-\lambda t}\int_0^t e^{(\lambda-\sigma)s}ds \\ &\leq a|\eta|e^{-\lambda t} + \frac{2a^2|\eta|l}{\sigma}e^{-\sigma t} \leq a|\eta|(1 + \frac{2al}{\sigma})e^{-\sigma t}. \end{aligned}$$

Similarly,

$$|z_m(t)| \leq \int_t^\infty 2a^2|\eta|le^{\lambda(t-s)}e^{-\sigma s}ds \leq \frac{2a^2|\eta|l}{3\sigma}e^{-\sigma t}.$$

Hence

$$\begin{aligned} |x_m(t)| &\leq |y_m(t)| + |z_m(t)| \\ &\leq a|\eta|(1 + \frac{8al}{3\sigma})e^{-\sigma t} \leq 2a|\eta|e^{-\sigma t}. \end{aligned}$$

We take into account here that (4.8) implies

$$\frac{8al}{3\sigma} = \frac{16al}{3\lambda} < 1.$$

Let us show now that for all m the inequality

$$|x_m(t) - x_{m-1}(t)| \leq 2a|\eta|\left(\frac{4al}{\lambda}\right)^{m-1} e^{-\sigma t} \tag{4.16}$$

holds. For $m = 1$ (4.16) is a consequence of (4.15). Suppose that (4.16) is valid and estimate $|x_{m+1}(t) - x_m(t)|$. We have

$$|y_{m+1}(t) - y_m(t)| \leq \int_0^t ae^{-\lambda(t-s)} l |x_m(s) - x_{m-1}(s)| ds$$
$$\leq 2la^2|\eta| \left(\frac{4al}{\lambda}\right)^{m-1} e^{-\lambda t} \int_0^t e^{\sigma s} ds \leq 2a|\eta| \left(\frac{al}{\sigma}\right)\left(\frac{4al}{\lambda}\right)^{m-1} e^{-\sigma t};$$
$$|z_{m+1}(t) - z_m(t)| \leq 2a|\eta|\left(\frac{al}{3\sigma}\right)\left(\frac{4al}{\lambda}\right)^{m-1} e^{-\sigma t}.$$

Hence,

$$|x_{m+1}(t) - x_m(t)| \leq 2a|\eta| \left(\frac{4al}{3\sigma}\right)\left(\frac{4al}{\lambda}\right)^{m-1} e^{-\sigma t}.$$

So we have that if (4.16) is valid for m, it is also valid for $m + 1$. It follows from (4.9) and (4.15) that for all m and for $t \geq 0$

$$(y_m(t), z_m(t)) \in U_\varepsilon.$$

As $4al < \lambda$, it follows from (4.16) that the sequence $x_m(t)$ converges, and the convergence is uniform with respect to $|\eta| < \Delta$ and to $t \geq 0$. Denote

$$\lim_{m\to\infty} x_m(t) = \tilde{x}(t, \eta) = \begin{pmatrix} \tilde{y}(t, \eta) \\ \tilde{z}(t, \eta) \end{pmatrix}.$$

The function $x_m(t)$ is continuous with respect to η. For $y_m(t)$ the continuity is evident, for $z_m(t)$ it follows from (4.7). Passing to the limit as $m \to \infty$ in (4.14), (4.15), we conclude that $\tilde{y}(t, \eta), \tilde{z}(t, \eta)$ is a solution of (4.12), satisfying (4.10), (4.11). So the map α defined by $\alpha(\eta) = \tilde{z}(0, \eta)$ for $|\eta| < \Delta$ has all properties described in the statement of the lemma. □

Consider the set

$$W^s_{loc}(0) = \{(y, z) : |y| < \Delta, z = \alpha(y)\}.$$

Lemma 4.2. *If $x \in W^s_{loc}(0)$ then there exists $t_0 \geq 0$ such that $\varphi(t, x) \in W^s_{loc}(0)$ for $t \geq t_0$.*

Proof. First let us show that for $\eta, |\eta| < \Delta$, system (4.12) cannot have two distinct solutions $y(t), z(t)$ such that $(y(t)), z(t)) \in U_\varepsilon$ for $t \geq 0$. To get a contradiction suppose that there are two solutions $x_1(t), x_2(t)$ having the described property. Let

$$x_i(t) = \begin{pmatrix} y_i(t) \\ z_i(t) \end{pmatrix}, \quad i = 1, 2.$$

We have then

$$\begin{aligned} y_1(t) - y_2(t) &= \int_0^t e^{A_1(t-s)}[F_1(x_1(s)) - F_1(x_2(s))]ds, \\ z_1(t) - z_2(t) &= -\int_t^\infty e^{A_2(t-s)}[F_2(x_1(s)) - F_2(x_2(s))]ds. \end{aligned} \tag{4.17}$$

It follows from (4.6), (4.7), (4.17) that

$$|y_1(t) - y_2(t)| \leq al \int_0^t e^{-\lambda(t-s)}|x_1(s) - x_2(s)|ds,$$

$$|z_1(t) - z_2(t)| \leq al \int_t^\infty e^{\lambda(t-s)}|x_1(s) - x_2(s)|ds.$$

Let

$$\mu = \sup_{t \geq 0} |x_1(t) - x_2(t)|,$$

then

$$|y_1(t) - y_2(t)| \leq \frac{al\mu}{\lambda}, |z_1(t) - z_2(t)| \leq \frac{al\mu}{\lambda}.$$

Consequently,

$$\mu \leq \frac{2al\mu}{\lambda.} \tag{4.18}$$

But as $2al < \lambda, \mu \geq 0$, (4.18) implies $\mu = 0$.

Let us show now that if $|\eta| < \Delta$ and $\tilde{z} \neq \alpha(\eta)$ then the solution $(y(t, \eta, \tilde{z}), z(t, \eta, \tilde{z}))$ leaves U_ε as t increases. Consider the corresponding solution

$$x(t) = \begin{pmatrix} y(t, \eta, \tilde{z}) \\ z(t, \eta, \tilde{z}) \end{pmatrix}.$$

To get a contradiction suppose that $x(t) \in U_\varepsilon$ for $t \geq 0$. Then the function $F_2(x(t))$ is bounded for $t \geq 0$. It follows from (4.7) that the integral

$$\int_t^\infty e^{A_2(t-s)} F_2(x(s))ds \tag{4.19}$$

converges and is bounded for $t \geq 0$. So $x(t)$ is a solution of the following system

$$y(t,\eta,\tilde{z}) = e^{A_1 t}\eta + \int_0^t e^{A_1(t-s)} F_1(x(s))ds,$$
$$z(t,\eta,\tilde{z}) = e^{A_2 t} z_0 - \int_t^\infty e^{A_2(t-s)} F_2(x(s))ds. \qquad (4.20)$$

We proved earlier the uniqueness of a solution of system (4.12) in U_ε. It follows from our assumption that $z_0 \neq 0$ (note that if $z_0 = 0$ then (4.12) and (4.20) coincide, so $\tilde{z} = \alpha(\eta)$). As

$$|z_0| = |e^{-A_2 t} e^{A_2 t} z_0| \leq a e^{-\lambda t} |e^{A_2 t} z_0|$$

for $t \geq 0$, we have

$$|e^{A_2 t} z_0| \geq \frac{e^{\lambda t}}{a} |z_0| \underset{t \to +\infty}{\longrightarrow} \infty.$$

The boundedness of (4.19) implies now that $|z(t,\eta,\tilde{z})|$ is unbounded as $t \to +\infty$, that leads us to a contradiction with the inclusion $x(t) \in U_\varepsilon$.

To finish the proof consider $x \in W^s_{loc}(0)$ and let

$$\varphi(t,x) = \begin{pmatrix} y(t) \\ z(t) \end{pmatrix}.$$

It follows from (4.10), (4.11) that there exists $t_0 \geq 0$ such that for $t \geq t_0$

$$|y(t)| < \Delta, \ |z(t)| < \varepsilon.$$

Using the established property of the map α we have that $z(t) = \alpha(y(t))$ for $t \geq t_0$, that means $\varphi(t,x) \in W^s_{loc}(0)$ for these t. □

Remark. It follows from (4.10), (4.11) and from the proof of Lemma 4.2 that if

$$x_0 = \begin{pmatrix} y_0 \\ z_0 \end{pmatrix} \in W^s_{loc}(0)$$

and

$$2a|y_0|e^{-\sigma t_0} < \Delta$$

then $\varphi(t,x_0) \in W^s_{loc}(0)$ for $t \geq t_0$. So if

$$|y_0| < \Delta_0 = \frac{\Delta}{2a}$$

then $\varphi(t,x_0) \in W^s_{\text{loc}}(0)$ for all $t \geq 0$.

To prove that the map α is of class C^1 we show now that α has the following property.

Lemma 4.3. *Let*

$$\tilde{x}(t,\eta) = \begin{pmatrix} \tilde{y}(t,\eta) \\ \tilde{z}(t,\eta) \end{pmatrix}$$

be a solution of (4.12), $|\eta| < \Delta$. Then for η_1, η_2 such that $|\eta_1| < \Delta, |\eta_2| < \Delta$ we have

$$|\tilde{x}(t,\eta_1) - \tilde{x}(t,\eta_2)| \leq 2a|\eta_1 - \eta_2|e^{-\sigma t}, \quad t \geq 0, \tag{4.21}$$

$$|\tilde{z}(t,\eta_1) - \tilde{z}(t,\eta_2)| \leq \frac{4a^2 l}{3\lambda}|\eta_1 - \eta_2|e^{-\sigma t}, \quad t \geq 0. \tag{4.22}$$

Proof. Consider successive approximations

$$x_m(t) = \begin{pmatrix} y_m(t,\eta) \\ z_m(t,\eta) \end{pmatrix}$$

constructed for the proof of lemma 4.1. We show that analogues of (4.21), (4.22) are valid for $x_m(t,\eta), z_m(t,\eta)$. For $m = 0$ it is evidently true. Suppose that our statement is true for $m-1$, then

$$|y_m(t,\eta_1) - y_m(t,\eta_2)|$$

$$\leq |e^{A_1 t}(\eta_1 - \eta_2)| + |\int_0^t e^{A_1(t-s)}[F_1(x_{m-1}(s,\eta_1)) - F_1(x_{m-1}(s,\eta_2))]ds|$$

$$\leq ae^{-\lambda t}|\eta_1 - \eta_2| + \int_0^t 2a^2 l e^{-\lambda(t-s)}|\eta_1 - \eta_2|e^{-\sigma s}ds|$$

$$\leq ae^{-\sigma t}(|\eta_1 - \eta_2| + \frac{2al}{\sigma}|\eta_1 - \eta_2|) = ae^{-\sigma t}|\eta_1 - \eta_2|\left(1 + \frac{4al}{\lambda}\right),$$

$$\begin{aligned}
&|z_m(t,\eta_1) - z_m(t,\eta_2)| \\
&\le \left|\int_t^\infty e^{A_2(t-s)}[F_2(x_{m-1}(s,\eta_1)) - F_2(x_{m-1}(s,\eta_2))]ds\right| \\
&\le \int_t^\infty 2a^2 l e^{\lambda(t-s)}|\eta_1 - \eta_2|e^{-\sigma s}ds \\
&\le \frac{2a^2 l}{3\sigma}|\eta_1 - \eta_2|e^{-\sigma t} = \frac{4a^2 l}{3\lambda}|\eta_1 - \eta_2|e^{-\sigma t}.
\end{aligned}$$

We have established (4.22) for z_m. The estimate (4.21) for x_m follows from the inequality

$$\begin{aligned}
|x_m(t,\eta_1) - x_m(t,\eta_2)| &\le |y_m(t,\eta_1) - y_m(t,n_2)| \\
&+ |z_m(t,\eta_1) - z_m(t,\eta_2)| \\
&\le ae^{-\sigma t}|\eta_1 - \eta_2|\left(1 + \frac{4al}{\lambda} + \frac{4al}{3\sigma}\right)
\end{aligned}$$

(we take into account that $16al < \lambda$). As the sequence $x_m(t,\eta)$ converges to $x(t,\eta)$ uniformly with respect to t,η, we obtain (4.21), (4.22) passing to limit as $m \to \infty$ in corresponding inequalities for x_m, z_m.

Corollary. *As $\alpha(\eta) = \tilde{z}(0,\eta)$, it follows from (4.22) that $4a^2l/3\lambda$ is Lipschitz constant for the map α on the domain $|\eta| < \Delta$.*

Lemma 4.4. *The map α is of class C^1 for $|\eta| < \Delta$ and $\frac{\partial\alpha}{\partial\eta}(0) = 0$.*

Proof. Let e_i be the ith vector of the standard basis of $\mathbb{R}^{n_1}$, $i = 1,\dots,n_1$. Consider $h \in \mathbb{R}$ and let $h_i = he_i$. Consider a solution $\tilde{x}(t,\eta)$ of (4.12) and let

$$p(t,\eta,h_i) = \tilde{x}(t,\eta + h_i) - \tilde{x}(t,\eta).$$

With respect to the decomposition $x = (y,z)$ we let

$$p(t,\eta,h_i) = \begin{pmatrix} q(t,\eta,h_i) \\ r(t,\eta,h_i) \end{pmatrix}.$$

Then we have

$$q(t,\eta,h_i) = e^{A_1 t}h_i + \int_0^t e^{A_1(t-s)}[F_1(\tilde{x}(s,\eta+h_i)) - F_1(\tilde{x}(s,\eta))]ds,$$

$$r(t,\eta,h_i) = -\int_t^\infty e^{A_2(t-s)}[F_2(\tilde{x}(s,\eta+h_i)) - F_2(\tilde{x}(s,\eta))]ds.$$

As F_1, F_2 are of class C^1 we can set

$$q(t,\eta,h_i) = e^{A_1 t}h_i + \int_0^t e^{A_1(t-s)}\left[\frac{\partial F_1}{\partial x}p(t,\eta,h_i) + \Delta_1\right]ds,$$

$$r(t,\eta,h_i) = -\int_t^\infty e^{A_2(t-s)}\left[\frac{\partial F_2}{\partial x}p(t,\eta,h_i) + \Delta_2\right]ds, \qquad (4.23)$$

here

$$\frac{|\Delta_1|}{|p|} \to 0, \ \frac{|\Delta_2|}{|p|} \to 0 \ \text{as} \ h \to 0.$$

It follows from (4.21) that

$$|p(t,\eta,h_i)| \leq 2a|h|e^{-\sigma t}, \ \ t \geq 0. \qquad (4.24)$$

The matrices $\frac{\partial F_1}{\partial x}, \frac{\partial F_2}{\partial x}$ are uniformly continuous on the compact U_ε. Therefore there is a function $\nu(h)$ such that $\nu(h) \to 0$ as $h \to 0$ and

$$|\Delta_j| \leq \nu(h)|h|e^{-\sigma t} \ \ \text{for}\, t \geq 0, \ j = 1, 2.$$

Fix $\mu \in \mathbb{R}$, $|\mu| < \Delta$, and consider the system of integral equations

$$\tilde{\gamma}_1(t,\eta) = e^{A_1 t}\mu e_i + \int_0^t e^{A_1(t-s)}\frac{\partial F_1}{\partial x}(\tilde{x}(s,\eta))\tilde{\gamma}ds,$$

$$\tilde{\gamma}_2(t,\eta) = -\int_t^\infty e^{A_2(t-s)}\frac{\partial F_2}{\partial x}(\tilde{x}(s,\eta))\tilde{\gamma}ds, \qquad (4.25)$$

where

$$\tilde{\gamma}(t,\eta) = \begin{pmatrix} \tilde{\gamma}_1(t,\eta) \\ \tilde{\gamma}_2(t,\eta) \end{pmatrix}$$

according to the decomposition $\mathbb{R} = \mathbb{R}^{n_1} \times \mathbb{R}^{n_2}$.

System (4.25) is of the same form as (4.12). To prove the existence of a solution for (4.12) we used the estimates (4.6), (4.7) of the norms $\|e^{A_1 t}\|$, $\|e^{A_2 t}\|$, the estimate $|\eta| < \Delta$ of the multiplier of $e^{A_1 t}$ in the first equation of (4.12), and also the estimate of the Lipschitz constant of F_1, F_2.

The multipliers of $e^{A_1(t-s)}$, $e^{A_2(t-s)}$ in integrands in (4.25) are linear with respect to γ. For $x \in U_\varepsilon$ we have estimates

$$\left\|\frac{\partial F_1}{\partial x}\right\|, \left\|\frac{\partial F_2}{\partial x}\right\| < l_0.$$

Therefore to prove the existence of a solution for (4.25) we have to repeat the proof of the existence for the solution of (4.12). The function

$$\gamma(t,\eta) = \frac{\tilde{\gamma}(t,\eta)}{\mu}$$

is a solution of the system

$$\begin{aligned}\gamma_1(t,\eta) &= e^{A_1 t}e_i + \int_0^t e^{A_1(t-s)}\frac{\partial F_1}{\partial x}(\tilde{x}(t,\eta))\gamma ds,\\ \gamma_2(t,\eta) &= -\int_t^\infty e^{A_2(t-s)}\frac{\partial F_2}{\partial x}(\tilde{x}(t,\eta))\gamma ds.\end{aligned} \tag{4.26}$$

Divide both inequalities in (4.23) by h and subtract the equalities in (4.26):

$$\begin{aligned}\frac{q}{h} - \gamma_1 &= \int_0^t e^{A_1(t-s)}\left[\frac{\partial F_1}{\partial x}\left(\frac{p}{h} - \gamma\right) + \frac{\Delta_1}{h}\right] ds,\\ \frac{r}{h} - \gamma_2 &= -\int_t^\infty e^{A_2(t-s)}\left[\frac{\partial F_2}{\partial x}\left(\frac{p}{h} - \gamma\right) + \frac{\Delta_2}{h}\right] ds.\end{aligned} \tag{4.27}$$

Using estimates (4.6), (4.7) we obtain

$$\begin{aligned}\left|\frac{q}{h} - \gamma_1\right| &\le a\int_0^t e^{-\lambda(t-s)}\left(\left|\frac{p}{h} - \gamma\right|l + \left|\frac{\Delta_1}{h}\right|\right) ds,\\ \left|\frac{r}{h} - \gamma_2\right| &\le a\int_t^\infty e^{\lambda(t-s)}\left(\left|\frac{p}{h} - \gamma\right|l + \left|\frac{\Delta_2}{h}\right|\right) ds.\end{aligned} \tag{4.28}$$

Let $\theta = \sup_{t\ge 0}|\frac{p}{h} - \gamma|$. It follows from (4.28) that

$$\begin{aligned}\theta &\le a[l\theta + \nu(h)]\left(\int_0^t e^{-\lambda(t-s)}ds + \int_t^\infty e^{\lambda(t-s)}ds\right)\\ &\le \frac{2al\theta}{\lambda} + \frac{2a\nu(h)}{\lambda}.\end{aligned} \tag{4.29}$$

As $2al < \lambda$ by (4.8), we obtain from (4.29) that $\theta \to 0$ as $h \to 0$. That implies the existence of

$$\frac{\partial x(t,\eta)}{\partial \eta_i} = \gamma(t,\eta)$$

(here η_i is the ith component of η).

It follows from the Corollary of Lemma 4.3 that

$$\left\|\frac{\partial \alpha}{\partial \eta}\right\| < \frac{4a^2 l}{3\lambda} \tag{4.30}$$

where l is the Lipschitz constant of F_1, F_2 at U_ε. Evidently $l \to 0$ as $\varepsilon \to 0$, so we obtain from (4.30) that

$$\frac{\partial \alpha}{\partial \eta}(0) = 0. \qquad \square$$

To prove the existence of a map β having all the properties described in the statement of Theorem 4.1, change t for $-t$. That completes the proof.

We call the smooth discs $W^s_{loc}(0), W^u_{loc}(0)$ constructed during the proof of Theorem 4.1 the local stable manifold of the rest point $x = 0$ and the local unstable manifold of the rest point $x = 0$ respectively.

A proof of the following theorem can be found in [24], and we omit this proof here.

Theorem 4.2. *If F_1, F_2 are of class C^r, $1 \le r \le \infty$, or $r = w$, then the maps α, β are also of class C^r.*

2. There is a convenient change of variables such that with respect to new variables the discs $W^s_{loc}(0), W^u_{loc}(0)$ are balls in coordinate hyperplanes.

Consider new variables

$$\xi = y - \beta(z), \quad \eta = z - \alpha(y). \tag{4.31}$$

It follows from (4.4) and (4.5) that the Jacobi matrix at the origin is

$$\frac{\partial(\xi,\eta)}{\partial(y,z)}(0,0) = \begin{pmatrix} E_{n_1} & 0 \\ 0 & E_{n_2} \end{pmatrix} = E.$$

If F_1, F_2 are of class C^r then by the Inverse Function Theorem there is a diffeomorphism H of class C^r mapping some neighborhood of the origin onto itself and such that

$$(y,z) = H(\xi,\eta) = (H_1(\xi,\eta), H_2(\xi,\eta)),$$
$$H_1(0,0) = 0, H_2(0,0) = 0, \frac{\partial H}{\partial(\xi,\eta)}(0,0) = E. \tag{4.32}$$

It follows from (4.32) that

$$H_1(\xi,\eta) = \xi + h_1(\xi,\eta), H_2(\xi,\eta) = \eta + h_2(\xi,\eta),$$

where

$$h_i(0,0) = 0, \frac{\partial h_i}{\partial(\xi,\eta)}(0,0) = 0, \quad i = 1,2.$$

Let us differentiate the first equality in (4.31) by t taking (4.2) into account:

$$\begin{aligned}\dot{\xi} &= \dot{y} - \frac{\partial \beta}{\partial z}\dot{z} \\ &= A_1 y + F_1(y,z) - \frac{\partial \beta}{\partial z}[A_2 z + F_2(y,z)] \\ &= A_1\xi + \widetilde{Y}(\xi,\eta). \end{aligned} \tag{4.33}$$

It follows from properties of F_1, F_2, β that

$$\frac{|\widetilde{Y}(\xi,\eta)|}{|\xi|+|\eta|}\underset{|\xi|+|\eta|\to 0}{\longrightarrow} 0. \tag{4.34}$$

Note that $\widetilde{Y}$ is of class C^{r-1} (as $\frac{\partial\beta}{\partial z}$ is of class C^{r-1}). Similarly we have

$$\dot{\eta} = A_2\eta + \widetilde{Z}(\xi,\eta), \tag{4.35}$$

$\widetilde{Z}$ has properties analogous to properties of $\widetilde{Y}$.

We noted in the remark after Lemma 4.2 that if

$$x_0 = \begin{pmatrix} y_0 \\ z_0 \end{pmatrix} \in W^s_{loc}(0)$$

and $|y_0| < \Delta_0$ then $\varphi(t,x_0) \in W^s_{loc}(0)$ for $t \geq 0$. Therefore for a point from $\{\eta = 0\} \cap H^{-1}(\{|y| < \Delta_0,\ z = \alpha(y)\})$ its positive trajectory remains in the hyperplane $\{\eta = 0\}$. From this we conclude that for small $|\xi|$

$$0 = A_2 0 + \widetilde{Z}(\xi, 0)$$

(we substitute $\eta = 0$ into (4.35)), so $\widetilde{Z}(\xi,0) = 0$. Similarly $\widetilde{Y}(0,\eta) = 0$ for small $|\eta|$. It follows from the Hadamard Lemma that there is an

$n_1 \times n_1$ matrix Y and $n_2 \times n_2$ matrix Z such that in some neighborhood of the origin Y, Z are of class C^{r-2} and

$$\widetilde{Y}(\xi,\eta) = Y(\xi,\eta)\xi, \ \widetilde{Z}(\xi,\eta) = Z(\xi,\eta)\eta.$$

Let us suppose that F_1, F_2 are of class C^2 in (4.2). In this case we can reduce the system (4.33), (4.35) to the form

$$\begin{aligned} \dot{\xi} &= (A_1 + Y(\xi,\eta))\xi, \\ \dot{\eta} &= (A_2 + Z(\xi,\eta))\eta \end{aligned} \tag{4.36}$$

in some neighborhood U_0 of the origin. Here Y, Z are continuous matrices, and

$$Y(0,0) = 0, Z(0,0) = 0. \tag{4.37}$$

For the rest point $(\xi,\eta) = (0,0)$ of (4.36) we have

$$\begin{aligned} W^s_{loc}(0) &= \{(\xi,\eta) : |\xi| < \Delta_1, \eta = 0\}, \\ W^u_{loc}(0) &= \{(\xi,\eta) : \xi = 0, |\eta| < \Delta_1\} \end{aligned}$$

where $\Delta_1 > 0$ is small enough.

The change of variables (4.31) is local. We can construct corresponding global change of variables using standard smoothing procedures. We are going to use smoothing procedures in various situations below. We describe the techniques we need in the following lemma.

Lemma 4.5. *Let g be a scalar function of class C^r, $1 \leq r \leq \infty$, in a neighborhood V of the origin of $\mathbb{R}^n$. Suppose that*

$$g(0) = 0, \frac{\partial g}{\partial x}(0) = 0. \tag{4.38}$$

Then for each $\varepsilon > 0$ there is a subneighborhood V_0 of V and a function $\tilde{g}$ of class C^r defined on $\mathbb{R}^n$ such that:

(1) $\tilde{g}(x) = g(x)$ for $x \in V_0$;

(2) $|\frac{\partial \tilde{g}}{\partial x_i}(x)| < \varepsilon$ for $x \in \mathbb{R}^n$, $i = 1, \dots, n$.

Proof. Consider a scalar function $\eta(t)$ of class C^∞ on $\mathbb{R}$ having the following properties:

(1) $\eta(t) = 1$ for $t \le 1$;
(2) $\eta(t) = 0$ for $t \ge 2$;
(3) $0 < \eta(t) < 1$ for $t \in (1,2)$;
(4) $-2 \le \eta'(t) \le 0$ for $t \in \mathbb{R}$.

See Figure 6 for the graph of $\eta(t)$. We leave it to the reader to find a formula defining a function $\eta(t)$ having the properties (1)–(4).

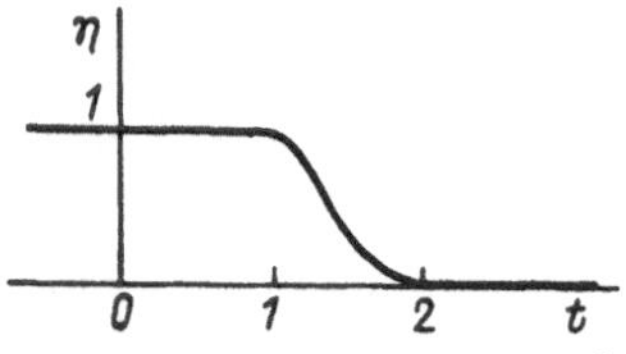

Figure 6.

Fix $\delta > 0$ such that the ball $x^2 < 2\delta$ is a subset of V (as usual, $x^2 = |x|^2 = x_1^2 + \cdots + x_n^2$). Consider the function $g_\delta(x)$ defined by

$$g_\delta(x) = g(x)\eta(\frac{x^2}{\delta}) \text{ for } x^2 < 2\delta,$$
$$g_\delta(x) = 0 \text{ for } x^2 \ge 2\delta.$$

Evidently g_δ is of class C^r in $\mathbb{R}^n$ and $g_\delta(x) = g(x)$ for $x^2 < \delta$. Fix $\varepsilon > 0$. We claim that for small δ we have

$$\left|\frac{\partial g_\delta(x)}{\partial x_i}\right| < \varepsilon, \ i = 1, \ldots, n, \tag{4.40}$$

for $x \in \mathbb{R}^n$. So we can take g_δ with small δ as $\tilde{g}$.

For $x^2 \ge 2\delta$ we have $\frac{\partial g_\delta}{\partial x} = 0$, so it remains to obtain (4.40) for $x^2 < 2\delta$. Consider

$$\frac{\partial g_\delta(x)}{\partial x_i} = \frac{\partial g(x)}{\partial x_i}\eta\left(\frac{x^2}{\delta}\right) + g(x)\eta'\left(\frac{x^2}{\delta}\right)\frac{2x_i}{\delta}.$$

Find $\delta_1 > 0$ such that for $x^2 < 2\delta_1$

$$\left|\frac{\partial g(x)}{\partial x_i}\right| < \frac{\varepsilon}{2}.$$

To estimate the second term we state it in the following way:

$$|g\eta'\frac{2x_i}{\delta}| = \frac{|g(x)|}{|x|}|\eta'|\frac{2|x_i|\,|x|}{\delta}.$$

From (4.38) it follows that there is $\delta_2 > 0$ such that for $0 < x^2 < 2\delta_2$

$$\frac{|g(x)|}{|x|} < \frac{\varepsilon}{16}.$$

Evidently, $|x_i| \le x$, so we have $2|x_i| \cdot |x| \le 2x^2$. Taking into account that $|\eta'| \le 2$ we have for $0 < x^2 < 2\delta$, where $\delta \le \delta_2$, that

$$\frac{|g(x)|}{|x|}|\eta'|\frac{2|x_i|\,|x|}{\delta} < \frac{\varepsilon}{2}.$$

Finally take $\delta = \min(\delta_1, \delta_2)$. □

To turn the local change of variables (4.31) into a global one, consider the function $\eta(t)$ we used in the proof of Lemma 4.5 and $\delta > 0$, and define new variables ξ, η by

$$\xi = \begin{cases} y - \beta(z)\eta\left(\frac{y^2+z^2}{\delta}\right), & y^2+z^2 \le 2\delta, \\ y, y^2+z^2 > 2\delta, \end{cases} \tag{4.41}$$

$$\eta = \begin{cases} z - \alpha(y)\eta\left(\frac{y^2+z^2}{\delta}\right), & y^2+z^2 \le 2\delta, \\ z, y^2+z^2 > 2\delta. \end{cases} \tag{4.42}$$

It follows from the proof of Lemma 4.5 that for any $\varepsilon > 0$ we can find $\delta > 0$ such that inequality

$$\left\| E_n - \frac{\partial(\xi,\eta)}{\partial(y,z)} \right\| < \varepsilon$$

holds for all y, z. It is easy to see that if ε is small enough, then (4.41), (4.42) defines a diffeomorphism which maps $\mathbf{R}^n$ onto $\mathbf{R}^n$ and coincides with (4.31) in a neighborhood of the origin.

Let us study the structure of trajectories of (4.36) in a neighborhood of the rest point $x = 0$. Consider a solution $(\xi(t), \eta(t))$ of (4.36).

We know that for the eigenvalues λ_j of the matrix A_1 we have $Re\lambda_j < 0$, and for eigenvalues λ_j of A_2 we have $Re\lambda_j > 0$. Fix $\mu_1 > 0$ such that $Re\lambda_j < -2\mu_1$ for the eigenvalues of A_1. It follows from the Canonical Form Theorem that there is a nonsingular change of variables $\xi = Su$ having the following property:

$$S^{-1}A_1S = J_0 + J_1,$$

where $\|J_1\| < \mu_1$ and J_0 is block-diagonal,

$$J_0 = \mathrm{diag}\left(\lambda_1, \dots, \lambda_k, \begin{pmatrix} a_1 & -b_1 \\ b_1 & a_1 \end{pmatrix}, \dots, \begin{pmatrix} a_m & -b_m \\ b_m & a_m \end{pmatrix}\right).$$

Here $\lambda_1, \dots, \lambda_k$ are real eigenvalues of A_1, and $a_1 \pm b_1 i, \dots, a_m \pm b_m i$ are complex eigenvalues of A_1.

With respect to the change of variables $\xi = Su$ the system

$$\dot{\xi} = A_1\xi$$

reduces to

$$\dot{u} = S^{-1}A_1Su. \tag{4.43}$$

Let us estimate the form $< S^{-1}A_1Su, u >$:

$$\begin{aligned} < S^{-1}A_1Su, u > &=< J_0u, u > + < J_1u, u > \\ &= \lambda_1 u_1^2 + \dots + \lambda_k u_k^2 + a_1(u_{k+1}^2 + u_{k+2}^2) + \dots + \\ &\quad + a_m(u_{n_1-1}^2 + u_{n_1}^2) + < J_1u, u > \\ &\le -2\mu_1 u^2 + \mu_1 u^2 = -\mu_1 u^2. \end{aligned}$$

Consequently, if we differentiate the function u^2 with respect to system (4.43), we obtain

$$\frac{d}{dt}u^2 = 2 < S^{-1}A_1Su, u >\le -2\mu_1 u^2.$$

Suppose we have produced the change of variables $\xi = Su$ and an analogous change of variables $\eta = Tv$. For simplicity we denote new coordinates by ξ, η, and new matrices by A_1, A_2. Then we have

$$< A_1\xi, \xi >\le -\mu_1\xi^2, < A_2\eta, \eta > \ge \mu_2^2\eta^2 \tag{4.44}$$

for some $\mu_2 > 0$. Let $\mu = \frac{1}{2}\min(\mu_1, \mu_2)$.

Lemma 4.6. *There is a neighborhood U_1 of the origin having the following property: if $(\xi(t), \eta(t)) \in U_1$ for $t \in (t_1, t_2)$, then for any $t, \tau \in (t_1, t_2)$ such that $t \le \tau$ we have*

$$|\xi(\tau)| \le e^{-\mu(\tau-t)}|\xi(t)|, \tag{4.45}$$

$$|\eta(\tau)| \ge e^{\mu(\tau-t)}|\eta(t)|. \tag{4.46}$$

Proof. Choose a neighborhood U_1 of the origin such that for $(\xi, \eta) \in U_1$ the inequalities

$$| < Y(\xi,\eta)\xi, \xi > | \le \mu\xi^2, | < Z(\xi,\eta)\eta, \eta > | \le \mu\eta^2$$

hold. Such a neighborhood exists because Y, Z are continuous and satisfy (4.37). Consider the function $v(t) = \xi^2(t)$ and estimate its derivative with respect to system (4.36) in U_1:

$$\begin{aligned} \dot{v} &= 2 < \dot{\xi}, \xi >= 2\left[< A_1\xi, \xi > + < Y(\xi,\eta)\xi, \xi >\right] \\ &\le 2\left\{-2\mu\xi^2 + \mu\xi^2\right\} = -2\mu v. \end{aligned}$$

If we integrate the inequality $\frac{\dot{v}}{v} \le -2\mu$ from t to τ, we obtain

$$lnv(t) - lnv(t) \le -2\mu(\tau - t).$$

Consequently,

$$\xi^2(\tau) \le e^{-2\mu(\tau-t)}\xi^2(t)$$

and the last inequality is equivalent to (4.45). The proof of (4.46) is similar. □

It follows from Lemma 4.6 that while $(\xi(t), \eta(t)) \in U_1$, the norm of $\xi(t)$ exponentially decreases, and the norm of $\eta(t)$ exponentially increases as t grows. That means that the structure of trajectories of (4.36) in U_1 is similar to the structure of trajectories of the linear system

$$\dot{y} = A_1 y, \dot{z} = A_2 z \tag{4.47}$$

(see Figure 7). The last statement is refined by the following theorem (the Grobman-Hartman Theorem).

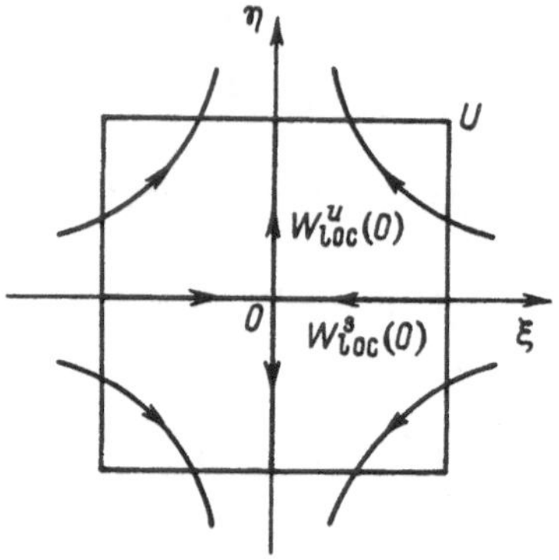

Figure 7.

Theorem 4.3. *The hyperbolic rest point* $(y, z) = (0, 0)$ *of system (4.2) is locally topologically conjugate to the rest point* $(y, z) = (0, 0)$ *of system (4.47).*

The Appendix of the book is devoted to a proof of the Grobman-Hartman Theorem.

The following important statement is an immediate consequence of Lemma 4.6.

Theorem 4.4. *There is a neighborhood* U *of a hyperbolic rest point* p *such that if a complete trajectory* γ *belongs to* U *then* $\gamma = p$.

Remark. To prove Lemma 4.6 and Theorem 4.4 we assumed that F is of class C^2 in (1.1). This was done to simplify proofs. Indeed these statements are valid if F is of class C^1, and we shall use them to study systems of class C^1.

3. We define now stable and unstable manifolds of a hyperbolic rest point p of system (1.1). The method we use is sometimes called the method of topological continuation.

Define the sets

$$W^s(p) = \{x \in \mathbf{R}^n : \varphi(t,x) \cap W^s_{loc}(p) \neq \emptyset\},$$
$$W^u(p) = \{x \in \mathbf{R}^n : \varphi(t,x) \cap W^u_{loc}(p) \neq \emptyset\}.$$

Sets $W^s(p)$ and $W^u(p)$ are called the stable manifold of p and the unstable manifold of p respectively. It follows immediately from the definition that sets $W^s(p)$ and $W^u(p)$ are invariant.

Lemma 4.7. *$x \in W^s(p)(x \in W^u(p)))$ if and only if $\varphi(t,x) \underset{t\to\infty}{\longrightarrow} p$ (respectively, $\varphi(t,x) \underset{t\to-\infty}{\longrightarrow} p$).*

Proof. Consider the case of $W^s(p)$. If $x \in W^s(p)$ then there is a τ such that $\varphi(\tau,x) \in W^s_{loc}(p)$. Then $\varphi(t,\varphi(\tau,x)) \to p$ as $t \to +\infty$. Consequently, $\varphi(t,x) \underset{t\to\infty}{\longrightarrow} p$.

Suppose now that $\varphi(t,x) \underset{t\to\infty}{\longrightarrow} p$. Then t_0 exists such that for $t \geq t_0$ the point $\varphi(t,x)$ belongs to the neighborhood U_1 of p described in Lemma 4.6. It follows from this lemma that $\varphi(t_0,x) \in W^s_{loc}(p)$. The case of $W^u(p)$ is considered similarly. □

Let us study the structure of $W^s(p)$ as a whole. We assume for simplicity that $p = 0$ and that with respect to coordinates ξ,η in $\mathbf{R}^n$, system (1.1) is in the form of (4.36) for small $|\xi|,|\eta|$. It was shown in Lemma 4.6 that there is a neighborhood U_1 of the origin such that $W^s_{loc}(0)$ is a smooth disc belonging to the hyperplane $\{\eta = 0\}$ and for the function $v = \xi^2$ we have in U_1

$$\dot{v} \leq -2\mu v,\ \mu > 0.$$

Consequently, if $a > 0$ is small enough then the $(n_1 - 1)$-dimensional sphere

$$\Sigma = \{(\xi,\eta) : \xi^2 = a, \eta = 0\}$$

belongs to $W^s_{loc}(0)$ and the trajectories on $W^s_{loc}(0)$ intersect Σ so that ξ^2 decreases on trajectories as t grows. It is easy to see that any trajectory on $W^s(0)$ different from the rest point $x = 0$ intersects Σ; for $x \in W^s(0)\setminus\{0\}$, $\varphi(t,x)$ leaves U_1 as $t \to -\infty$ and $\varphi(t,x) \to 0$ as $t \to +\infty$. As ξ^2 decreases along a trajectory $\varphi(t,x)$, $x \in W^s_{loc}(0)$, any trajectory from $W^s(0)\backslash\{0\}$ has a unique point of intersection with Σ. We say that Σ is a parametrizing sphere for $W^s(0)$. Consider a closed n_1-dimensional disc

$$\Sigma_0 = \{(\xi, \eta) : \xi^2 \leq a, \eta = 0\}$$

in $W^s_{loc}(0)$.

Consider the space $\mathbf{R}^{n_1}$ with coordinate w. Let

$$D = \{|w| \leq 1\} \subset \mathbf{R}^{n_1},\ S = \{|w| = 1\} \subset \mathbf{R}^{n_1}.$$

Define a map of $\mathbf{R}^{n_1}\backslash D$ onto $S \times (0, +\infty)$; take $w \in \mathbf{R}^{n_1}\backslash D$, consider the ray l beginning at $0 \in \mathbf{R}^{n_1}$ and containing w, and define by $\xi(w)$ the point of intersection of the ray l with S and by $\tau(w)$ the distance between w and $\xi(w)$. Evidently,

$$\xi(w) = \frac{w}{|w|},\ \tau(w) = |w - \frac{w}{|w|}|.$$

The map $w \mapsto (\xi(w), \tau(w))$ is a homeomorphism. Consider the number a from the definition of Σ. Define the map $b^s : \mathbf{R}^{n_1} \to W^s(0)$ by:

— for $w \in D\, b^s(w) = (\tilde{\xi}, 0)$, where $\tilde{\xi} = \sqrt{a}w$;

— for $w \in \mathbf{R}^{n_1}\backslash D\, b^s(w) = \varphi(-\tau(w), \sqrt{a}\xi(w))$.

It is easy to see that b^s maps $\mathbf{R}^{n_1}$ onto $W^s(0)$, and it is one-to-one. It follows from the continuity of φ that b^s is continuous. That means that b^s is a C^0 immersion of $\mathbf{R}^{n_1}$ into $\mathbf{R}^n$ (see Chapter 3 §6). We call the topology induced by the map b^s on W^s (0) the inner topology of $W^s(0)$.

Similarly a C^0 immersion $b^u : \mathbf{R}^{n_2} \to \mathbf{R}^n$ such that $b^u(\mathbf{R}^{n_2}) = W^u(0)$ is constructed.

Remarks. 1. The stable manifold of a hyperbolic rest point is not necessarily a submanifold of $\mathbf{R}^n$, i.e. the inner topology of $W^s(p)$ and the topology induced on $W^s(p)$ as on a subset of $\mathbf{R}^n$ may not coincide.

Consider the following system of differential equations on the plane $\mathbf{R}^2$ with the coordinates x, y:

$$\dot{x} = y, \dot{y} = x(1 - x^2). \tag{4.48}$$

System (4.48) has an integral $y^2 - x^2 + \frac{x^4}{2}$; consequently, the trajectories of (4.48) belong to curves

$$y = \pm\sqrt{c + x^2 - \frac{x^4}{2}}$$

(see Figure 8).

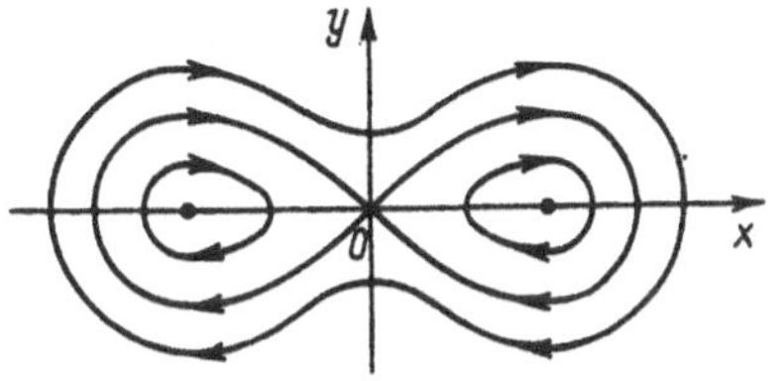

Figure 8.

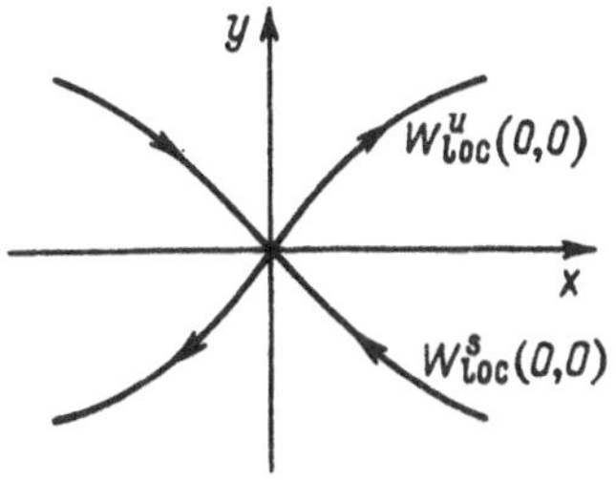

Figure 9.

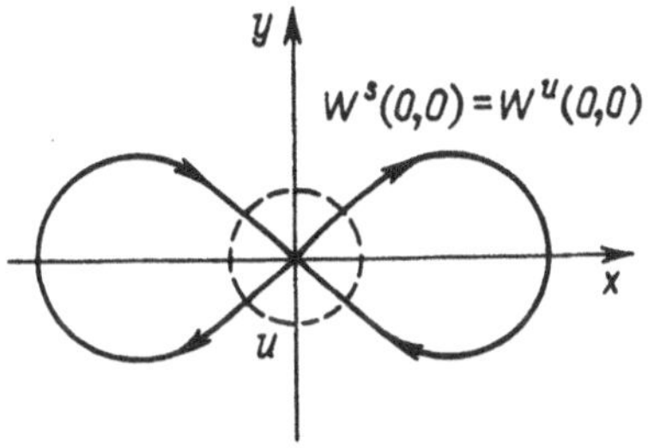

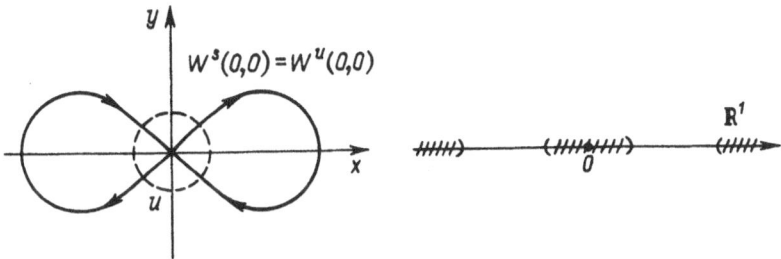

Figure 10.

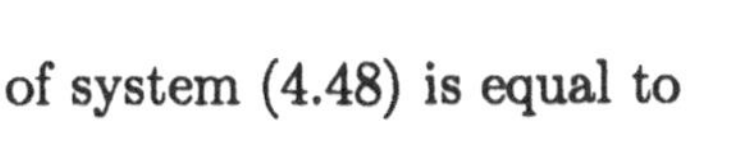

Figure 11.

At rest point (0,0) the Jacobi matrix of system (4.48) is equal to

$$\begin{pmatrix} 0 & 1 \\ 1 & 0 \end{pmatrix}$$

its eigenvalues are $\lambda_{1,2} = \pm 1$. So, (0,0) is a hyperbolic saddle rest point. See the local stable manifold and the local unstable manifold of (0,0) in Figure 9 and see the sets $W^s(0,0), W^u(0,0)$, in Figure 10.

Let b^s be an immersion $\mathbb{R}^1 \to \mathbb{R}^2$ such that $b^s(\mathbb{R}^1) = W^s(0,0)$. Consider a small neighborhood U of the origin (see the dotted circle in Figure 10). The set

$$(b^s)^{-1}(W^s(0,0) \cap U)$$

is shaded in Figure 11. It is easy to see that $W^s(0,0)$ is not a submanifold of $\mathbf{R}^2$.

2. If p is a hyperbolic saddle rest point and F is of class C^r, $r \geq 1$, for system (1.1) then there are immersions b^s and b^u of class C^r such that $b^s(\mathbf{R}^{n_1}) = W^s(p), b^u(\mathbf{R}^{n_2}) = W^u(p)$. For the immersion b^s and for a compact $K \subset \mathbf{R}^{n_1}$ the restriction of b^s on K is an embedding of class C^r mapping K into $\mathbf{R}^n$ [37]. In this book we consider closed balls $\{|w| \leq C\} \subset \mathbf{R}^{n_1}$ as such compacts.

3. The considerations in Sections 1,2 are local, we can use them not only for systems in $\mathbf{R}^n$ but also for systems on manifolds. We can also repeat the construction of Section 3 when the phase space is a manifold.

4. If p is a hyperbolic saddle point of an autonomous system in $\mathbf{R}^n$ or on an n-dimensional manifold M then

$$\dim W^s(p) + \dim W^u(p) = n. \tag{4.49}$$

Chapter 5

Periodic Point and Closed Trajectory

1. Consider a diffeomorphism $f : \mathbf{R}^n \to \mathbf{R}^n$, and let p be a periodic point of period k. Let us begin with the case $k = 1$, i.e. the case of a fixed point p.

We say that the fixed point p is hyperbolic if the eigenvalues λ_j of its derivative $Df(p)$ satisfy the following condition:

$$|\lambda_j| \neq 1 \tag{5.1}$$

Let A be the matrix of $Df(p)$. Consider a decomposition $\mathbf{R}^n = \mathbf{R}^{n_1} \times \mathbf{R}^{n_2}$ with coordinates y in $\mathbf{R}^{n_1}$, z in $\mathbf{R}^{n_2}$ such that according to this decomposition the matrix A is block-diagonal,

$$A = \begin{pmatrix} A_1 & 0 \\ 0 & A_2 \end{pmatrix}.$$

In this case A_1 is an $n_1 \times n_1$ matrix with eigenvalues $\lambda_1, \ldots, \lambda_{n_1}$, and $|\lambda_j| < 1$, $j = 1, \ldots, n_1$; A_2 is an $n_2 \times n_2$ matrix with eigenvalues $\lambda_{n_1+1}, \ldots, \lambda_n$, and $|\lambda_j| > 1$, $j = n_1 + 1, \ldots, n$. For simplicity we suppose that $p = 0$. Then in a neighborhood of the origin we can write

$$f(y, z) = (A_1 y + f_1(y, z),\ A_2 z + f_2(y, z)). \tag{5.2}$$

Here

$$f_i(0,0) = 0,\ \frac{\partial f_i}{\partial(y, z)}(0,0) = 0,\ \ i = 1, 2.$$

Using the same method as in Theorem 4.1, we can establish the following statement (the Stable Manifold Theorem for diffeomorphisms).

Theorem 5.1. *Let $x = 0$ be a hyperbolic fixed point of a diffeomorphism f of class C^1 having the form* (5.2) *in a neighborhood of the origin. Then there exist $\Delta > 0$ and maps*

$$\alpha : \{|y| < \Delta\} \to \mathbf{R}^{n_2},$$
$$\beta : \{|z| < \Delta\} \to \mathbf{R}^{n_1}$$

of class C^1 such that (4.4), (4.5) *hold and*

(1) *if x belongs to the set*

$$W^s_{loc}(0) = \{(y, z) : |y| < \Delta,\ z = \alpha(y)\}$$

then $f^k(x) \in W^s_{loc}(0)$ for $k \geq 0$ and $f^k(x) \underset{k\to+\infty}{\longrightarrow} 0$;

(2) *if x belongs to the set*

$$W^u_{loc}(0) = \{(y, z) : y = \beta(z), |z| < \alpha\}$$

then $f^k(x) \in W^u_{loc}(0)$ for $k \leq 0$ and $f^k(x) \underset{k\to-\infty}{\longrightarrow} 0$.

We call the smooth discs $W^s_{loc}(0)$ and $W^u_{loc}(0)$ the local stable manifold of the fixed point $x = 0$ and the local unstable manifold of the fixed point $x = 0$ respectively.

Let $n_2 = n - n_1$. If $n_1 = n, n_2 = 0$, $W^s_{loc}(0)$ is a neighborhood of the origin, and for any x in this neighborhood $f^k(x) \underset{k\to+\infty}{\longrightarrow} 0$. In this case we say that the fixed point $x = 0$ is attractive.

If $n_1 = 0, n_2 = n$, $W^u_{loc}(0)$ is a neighborhood of the origin, and for any x in this neighborhood $f^k(x) \underset{k\to-\infty}{\longrightarrow} 0$. In this case we say that the fixed point $x = 0$ is a repeller.

If $n_1 n_2 \neq 0$ we say that $x = 0$ is a hyperbolic saddle fixed point. Using considerations similar to the considerations of Section 2 in Chapter 4 we can show that there exists a neighborhood U_0 of the origin and $\mu \in (0, 1)$ such that with respect to some coordinates ξ, η we have the following:

(1) $W^s_{loc}(0) = \{\eta = 0\} \cap U_0$;

(2) $W^u_{loc}(0) = \{\xi = 0\} \cap U_0$;

(3) if for a point $(\xi_0, \eta_0) \in U_0$ $(\xi_k, \eta_k) = f^k(\xi_0, \eta_0) \in U_0$ for $k \in [k_1, k_2]$ then for any $k, \kappa \in [k_1, k_2]$ such that $k \leq \kappa$ the inequalities

$$|\xi_\kappa| \le \mu^{k-\kappa}|\xi_k|, \quad |\eta_\kappa| \ge \mu^{k-\kappa}|\eta_k|$$

hold.

So the structure of trajectories of the cascade generated by f in U_0 is analogous to the structure of trajectories of the cascade generated by the linear mapping $(\xi, \eta) \mapsto (\mu\xi, \eta/\mu)$.

For a hyperbolic fixed point p of a diffeomorphism f we define the stable and unstable manifolds:

$$W^s(p) = \{x \in \mathbf{R}^n : \{f^k(x)\} \cap W^s_{loc}(p) \neq \emptyset\},$$
$$W^u(p) = \{x \in \mathbf{R}^n : \{f^k(x)\} \cap W^u_{loc}(p) \neq \emptyset\}.$$

The manifolds $W^s(p), W^u(p)$ have properties analogous to properties of stable and unstable manifolds of hyperbolic rest points of autonomous systems of differential equations. In particular,

$$W^s(p) = \{x \in \mathbf{R}^n : f^k(x) \underset{k \to +\infty}{\longrightarrow} p\},$$
$$W^u(p) = \{x \in \mathbf{R}^n : f^k(x) \underset{k \to -\infty}{\longrightarrow} p\}$$

and there are immersions β^s, β^u of the same class C^r as the diffeomorphism f such that

$$\beta^s(\mathbf{R}^{n_1}) = W^s(p), \beta^u(\mathbf{R}^{n_2}) = W^u(p).$$

2. Next, consider a periodic point p of a diffeomorphism f such that the period k of p is greater than 1. In this case the orbit of p consists of k different points

$$p_0 = p,\ p_1 = f(p), \dots, p_{k-1} = f^{k-1}(p).$$

Each of the points $p_0, \dots, p_{k-1}$ is evidently a fixed point of the diffeomorphism f^k.

We say that the periodic point p is hyperbolic if p is a hyperbolic fixed point of the diffeomorphism f^k, i.e. if for the eigenvalues λ_j of $Df^k(p)$ inequalities (5.1) hold.

Consider two points p_i, p_j belonging to the orbit of p, $0 \le i < j \le k-1$. Denote $g = f^{j-i}$. Evidently

$$f^k = g^{-1} \circ f^k \circ g,$$

and it is easy to see that

$$Df^k(p_i) = Dg^{-1}(p_j) Df^k(p_j) Dg(p_i).$$

The matrices $Dg^{-1}(p_j), Dg(p_i)$ are inverse. Consequently the matrices $Df^k(p_i), Df^k(p_j)$ are conjugate and their sets of eigenvalues coincide. Thus, if one of the points $p_0, \dots, p_{k-1}$ is hyperbolic, then all the points $p_0, \dots, p_{k-1}$ are hyperbolic.

We denote by $\varphi(m, p)$ the trajectory of a point p in the cascade generated by f. For a point p_i, $i = 0, \dots, k-1$, we define $W^s(p_i)(W^u(p_i))$ as the stable (respectively, unstable) manifold of the fixed point p_i for f^k. It follows from this definition that

$$W^s(p_i) = \{x : f^{mk}(x) \underset{m \to +\infty}{\longrightarrow} p_i\},$$
$$W^u(p_i) = \{x : f^{mk}(x) \underset{m \to -\infty}{\longrightarrow} p_i\}.$$

If for the eigenvalues λ_j of the matrix $Df^k(p)$ $|\lambda_j| < 1$ for $1 \le j \le n_1$, and $|\lambda_j| > 1$ for $n_1 + 1 \le j \le n$, then $W^s(p_i), W^u(p_i)$ are images of immersions of $\mathbb{R}^{n_1}, \mathbb{R}^{n_2}$ respectively (note that $n_2 = n - n_1$), and the immersions are of the same class C^r as the diffeomorphism f.

We now define the stable and unstable manifolds of the trajectory $\varphi(m, p)$ by

$$W^s(\varphi(m,p)) = \bigcup_{0 \le i \le k-1} W^s(p_i),$$
$$W^u(\varphi(m,p)) = \bigcup_{0 \le i \le k-1} W^u(p_i).$$

Evidently $f(W^s(p_i)) = W^s(p_{i+1}), f(W^u(p_i)) = W^u(p_{i+1})$ (we define $p_k = p_0$), it follows that $W^s(\varphi(m,p))$, $W^u(\varphi(m,p))$ are invariant with respect to f.

If $i, j \in \{0, \dots, k-1\}$ and $i \ne j$ then $W^s(p_i) \cap W^s(p_j) = \emptyset$. To prove that, suppose that there exists $x \in W^s(p_i) \cap W^s(p_j)$. Then it follows from

$$f^{km}(x) \underset{m \to +\infty}{\longrightarrow} p_i, \ f^{km}(x) \underset{m \to +\infty}{\longrightarrow} p_j$$

that $p_i = p_j$, and that is impossible.

So, we see that $W^s(\varphi(m,p))$ is a union of k disjoint images of immersions of $\mathbf{R}^{n_1}$. In a similar way, we can describe the structure of $W^u(\varphi(m,p))$.

3. Consider an autonomous system of differential equations

$$\dot{x} = F(x). \tag{5.2}$$

For convenience we suppose in this section that $x \in \mathbf{R}^{n+1}$. Let φ be the flow generated by (5.3). Suppose that γ is a closed trajectory of (5.3) corresponding to a periodic solution $\psi(t)$. Denote by ω the least positive period of $\psi(t)$. Suppose that $0 \in \gamma$ and that $F(0) = (0, \dots, 0, a)$. Denote by S the coordinate hyperplane $\{x_{n+1} = 0\}$. We see that the vector $F(0)$ is orthogonal to S, so that it follows from Theorem 1.1 that there exists a diffeomorphism T mapping a neighborhood of the origin in S onto a neighborhood of the origin in S. It was shown in the proof of Theorem 1.1 that for $s = (x_1, \dots, x_n) \in S$ with small $|s|$, there are functions $t(s)$ and $\sigma(s) = (\tilde{x}_1(s), \dots, \tilde{x}_n(s))$ such that

$$t(0) = \omega,\ \sigma(0) = 0,\ T(s) = \sigma(s) = \varphi(t(s), (s, 0)) \cap S$$

and t, σ are of the same class C^r as the function F in (5.3).

We say that the closed trajectory γ is hyperbolic if $0 \in S$ is a hyperbolic fixed point of the diffeomorphism T, i.e. for the eigenvalues λ_j of $DT(0)$ inequalities (5.1) hold.

Denote by $\varphi_n(t, x)$, the vector of first n coordinates of the flow φ. Then we have

$$T(s) = \varphi_n(t(s), (s, 0)).$$

Therefore

$$DT(0) = \frac{\partial T}{\partial s}(0) = \frac{\partial \varphi_n}{\partial t} \cdot \frac{\partial t}{\partial s}(0) + \frac{\partial \varphi_n}{\partial s}(0).$$

It follows from our choice of $F(0)$ that

$$\frac{\partial \varphi_n}{\partial t}(0) = (0, \dots, 0),\ \frac{\partial T}{\partial s}(0) = \frac{\partial \varphi_n}{\partial s}(0).$$

So the matrix $DT(0)$ coincides with the matrix consisting of derivatives of the first n-components of the flow φ with respect to the first n-components s of initial values. It is known from the basic course of differential equations that if $\Phi(t)$ is the fundamental matrix of solutions of the system

$$\dot{y} = \frac{\partial F}{\partial x}(\psi(t))y, \quad y \in \mathbf{R}^{n+1}, \tag{5.4}$$

such that $\Phi(0) = E_{n+1}$, then

$$\frac{\partial \varphi}{\partial x}(0) = \Phi(\omega).$$

The eigenvalues of the matrix $\Phi(\omega)$ are usually called multipliers of the closed trajectory γ.

We claim that $\xi(t) = \dot{\psi}(t)$ is a solution of system (5.4). Let us differentiate the equality $\dot{\psi}(t) = F(\psi(t))$:

$$\dot{\xi} = \frac{\partial F}{\partial x}(\psi(t)) \cdot \dot{\psi}(t) = \frac{\partial F}{\partial x}(\psi(t))\xi.$$

It follows that $\xi(t) = \Phi(t)\xi(0)$. As

$$\xi(0) = \dot{\psi}(0) = F(0), \quad \xi(\omega) = F(\psi(\omega)) = F(0) = \xi(0),$$

we have

$$F(0) = \Phi(\omega)F(0).$$

That means that the matrix $\Phi(\omega)$ has eigenvalue 1, and that $F(0)$ is a corresponding eigenvector. We say that this eigenvalue is a standard multiplier of the closed trajectory γ.

As

$$\Phi(\omega)\begin{pmatrix} 0 \\ \vdots \\ 0 \\ a \end{pmatrix} = \begin{pmatrix} 0 \\ \vdots \\ 0 \\ a \end{pmatrix}$$

we have that

$$\frac{\partial \varphi}{\partial x}(0) = \Phi(\omega) = \begin{pmatrix} \frac{\partial \varphi_n}{\partial s}(0) & 0 \\ & \vdots \\ & 0 \\ \dots & 1 \end{pmatrix} = \begin{pmatrix} \frac{\partial T}{\partial s}(0) & 0 \\ & \vdots \\ & 0 \\ \dots & 1 \end{pmatrix}.$$

Therefore the eigenvalues of $\frac{\partial T}{\partial s}(0)$ are multipliers of γ different from the standard multiplier. So we obtain the following equivalent definition of the hyperbolicity of a closed trajectory γ: for the multipliers λ_j of γ different from the standard multiplier, inequalities (5.1) hold.

The hyperplane S chosen to define the Poincaré transformation T of the closed trajectory γ was taken in a special way. We are going to show that the eigenvalues of $DT(0)$ do not depend on the choice of a transversal S. Let S_1, S_2 be two transversals to the trajectory γ (see Figure 12), and let T_1, T_2 be the corresponding Poincaré transformations. Denoted by χ a diffeomorphism mapping a neighborhood of the origin on S_1 onto a neighborhood of the origin on S_2 (the existence of χ is established in Theorem 1.1). Consider $x \in S_1$, $x_1 = \chi(x) \in S_2, x_2 = T_1(x) \in S_1, x_3 = \chi(x_2) \in S_2$. It is easy to see that $x_3 = T_2(x_1)$, so that for $x \in S_1$, near $\gamma \cap S_1$, we have

$$\chi(T_1(x)) = T_2(\chi(x)).$$

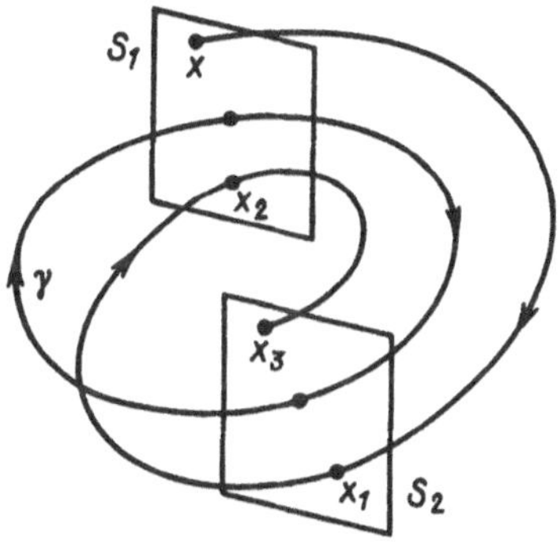

Figure 12

Hence, $T_1(x) = \chi^{-1}(T_2(\chi(x)))$, and

$$DT_1(0) = A^{-1} DT_2(0) A, \tag{5.5}$$

where $A = D\chi(0)$. It follows from (5.5) that the matrices $DT_1(0)$ and $DT_2(0)$ are conjugate. Consequently, their sets of eigenvalues coincide.

Let us apply results of Section 1 to the hyperbolic fixed point $0 \in S$ of the diffeomorphism T. Now suppose that for the eigenvalues $\lambda_1, \dots, \lambda_n$ of $DT(0)$

$$|\lambda_j| < 1,\ j = 1, \dots, n_1;\ |\lambda_j| > 1,\ j = n_1 + 1, \dots, n;\ n_2 = n - n_1.$$

It follows from Theorem 5.1 that there are smooth discs $W^s_{loc}(0)$, $W^u_{loc}(0)$ in S containing 0 and having the following properties:

– $\dim W^s_{loc}(0) = n_1$; for $x \in W^s_{loc}(0)$
$T^k(x) \in W^s_{loc}(0),\ k \geq 0;\quad T^k(x) \underset{k\to+\infty}{\longrightarrow} 0$;

– $\dim W^u_{loc}(0) = n_2$; for $x \in W^u_{loc}(0)$
$T^k(x) \in W^u_{loc}(0),\ k \leq 0;\quad T^k(x) \underset{k\to-\infty}{\longrightarrow} 0$.

There exists a neighborhood U of the origin in S such that if

$$x \in U \backslash (W^s_{\mathrm{loc}}(0) \cup W^u_{loc}(0))$$

then the trajectory $T^k(x)$ leaves U as $|k| \to \infty$.

Consider a point $\sigma \in W^s_{loc}(0)$ and the trajectory $\varphi(t, \sigma)$ of system (5.1). As $W^s_{loc}(0)$ is positively invariant with respect to T, after a rotation near γ, the trajectory $\varphi(t, \sigma)$ comes to W^s_{loc} at the point $\sigma_1 = T(\sigma)$ (see Figure 13).

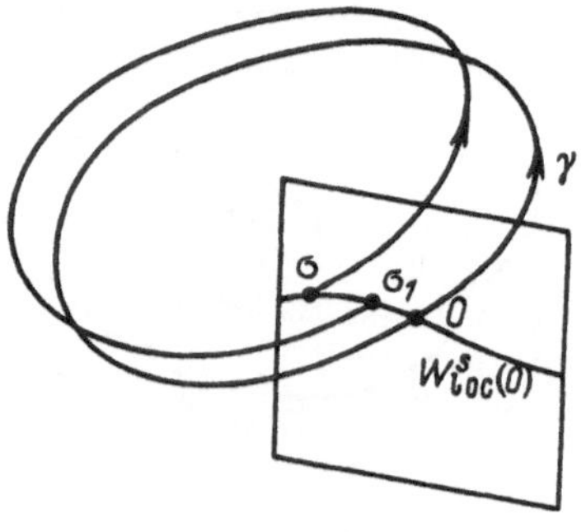

Figure 13

Consider the union of segments of trajectories $\varphi(t, \sigma)$ for $\sigma \in W^s_{loc}(0)$ (the ends of segments are points σ and $T(\sigma)$), this union is a $(n_1 + 1)$-dimensional manifold. We denote this manifold by $W^s_{loc}(\gamma)$ and call it the local stable manifold of the closed trajectory γ. Evidently the manifold $W^s_{loc}(\gamma)$ is positively invariant with respect to the flow φ, i.e. for $x \in W^s_{loc}(\gamma)$ and for $t \geq 0$ we have $\varphi(t, x) \in W^s_{loc}(\gamma)$.

Similarly, considering the union of segments of trajectories $\varphi(t,\sigma)$, $\sigma \in W^u_{loc}(0)$, such that the ends of segments are points σ and $T^{-1}(\sigma)$, we construct the local unstable manifold $W^u_{loc}(\gamma)$ of the trajectory γ. This manifold is negatively invariant with respect to φ.

Define the stable manifold of γ and the unstable manifold of γ by

$$W^s(\gamma) = \{x \in \mathbb{R}^{n+1} : \varphi(t,x) \cap W^s_{loc}(\gamma) \neq \emptyset\},$$
$$W^u(\gamma) = \{x \in \mathbb{R}^{n+1} : \varphi(t,x) \cap W^u_{loc}(\gamma \neq \emptyset\}.$$

From the properties of $W^s_{loc}(0)$, $W^u_{loc}(0)$ mentioned above and from the existence of a neighborhood U it follows that $x \in W^s(\gamma)(x \in W^u(\gamma))$ if and only if $\varphi(t,x)$ tends to γ as $t \to +\infty$ (respectively, $t \to -\infty$).

The equalities $n_1 + n_2 = n$, dim $W^s(\gamma) = n_1 + 1$, dim $W^u(\gamma) = n_2 + 1$ imply the equality

$$\dim W^s(\gamma) + \dim W^u(\gamma) = n + 2. \tag{5.6}$$

Considering the following example. Suppose that we study system (5.1) in $\mathbb{R}^3$ and that x, y, z are coordinates in $\mathbb{R}^3$. Suppose also that system (5.1) has a closed trajectory γ such that the plane $S : \{x = 0\}$ is transversal to γ and the corresponding Poincaré transformation $T : S \to S$ is given by

$$T(y,z) = (-\frac{y}{2}, -2z).$$

It is easy to see that $W^s_{loc}(0)$ in a neighborhood of the origin on the line $\{x = z = 0\}$. Fix a small directed segment of the line (this directed segment is the "thick vector" in Figure 14). It is shown in Figure 14 how the beginning and the end of this directed segment move along trajectories of the flow φ. Evidently the manifold $W^s_{loc}(\gamma)$ is a Möbius band with γ being a middle circle.

To describe the global structure of $W^s(\gamma), W^u(\gamma)$ we can repeat considerations of Section 3 in Chapter 4 (we leave it to the reader). The manifolds $W^s(\gamma), W^u(\gamma)$ are images of immersions of "cylinders" $\mathbb{R}^{n_1} \times S^1, \mathbb{R}^{n_2} \times S^1$ or of nonorientable fiberings over S^1 with fibers $\mathbb{R}^{n_1}, \mathbb{R}^{n_2}$ (see [27] for details).

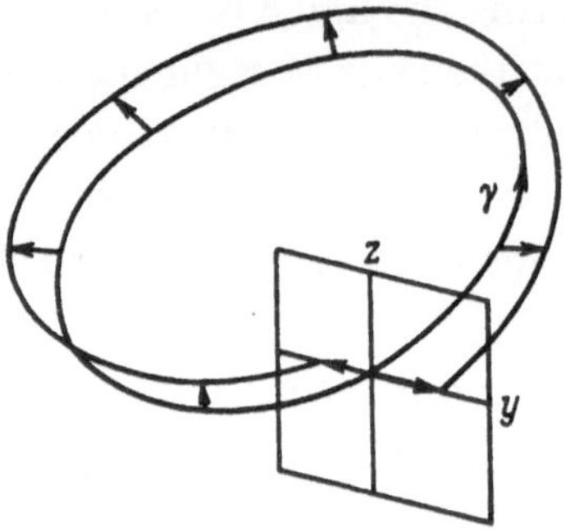

Figure 14

4. We can repeat all the considerations of this chapter in the case of flow and cascades on a smooth closed manifold M.

Chapter 6

Transversality

1. Let K, M be smooth manifolds, let L be a submanifold of M, and let f be a smooth map, $f : K \to M$.

We say that f is transversal to L at a point $x \in K$ if either $f(x) \notin L$ or

$$Df(x)(T_xK) + T_{f(x)}L = T_{f(x)}M. \tag{6.1}$$

The left side of the equality (6.1) is the sum of two linear subspaces of the space $T_{f(x)}M$: the image of T_xK by $Df(x)$, and the space $T_{f(x)}L$. Note that the following inequality is possible:

$$Df(x)(T_xK) \neq T_{f(x)}f(K).$$

Consider for example $K = \mathbb{R}$ with coordinate $t, M = \mathbb{R}^2$ with coordinates $x, y,\ L = \{y = 0\}, f(t) = (0, t^3)$. We leave it to the reader to see that

$$Df(0)(\mathbb{R}) \neq T_{(0,0)}(\{x = 0\}).$$

We say that the map f is transversal to L, if it is transversal to L at each point $x \in K$. We write $f \pitchfork L$ in this case.

Let, K, L be submanifolds of a manifold M. We say that K is transversal to L at $x_0 \in K$ if the inclusion map $i : K \to M$ is transversal to L at x_0. In other words, K is transversal to L at $x_0 \in K \cap L$ if

$$T_{x_0}K + T_{x_0}L = T_{x_0}M. \tag{6.2}$$

i.e.

$$T_{x_0}M = \{a + b : a \in T_{x_0}K,\ b \in T_{x_0}L\}.$$

We say that the submanifolds K, L of M are transversal if the inclusion map $i : K \to M$ is transversal to L. We write $K \pitchfork L$ in this case.

We can restate condition (6.2) in the following way:

$$\dim K + \dim L - \dim(T_{x_0}K \cap T_{x_0}L) = \dim M. \tag{6.3}$$

Note that $T_{x_0}K \cap T_{x_0}L$ in (6.3) is the intersection of two linear subspaces of the linear space $T_{x_0}M$.

To prove the equivalence of (6.2) and (6.3) note that (6.2) means the following: $T_{x_0}M$ contains $n = \dim M$ linearly independent vectors so that some of these vectors belong to $T_{x_0}K$, and the remaining vectors belong to $T_{x_0}L$. Let us count the maximal possible number of linearly independent vectors belonging to $T_{x_0}K$ and to $T_{x_0}L$ in the following way. Let dim $K = k$, dim $L = l$, dim $(T_{x_0}K \cap T_{x_0}L) = m$. Consider a basis B of $T_{x_0}K \cap T_{x_0}L$ consisting of m vectors. We can construct a basis B_1 of $T_{x_0}K$ adding $k - m$ linearly independent vectors to B. There exist $l - m$ linearly independent vectors in $T_{x_0}L$ such that these vectors are linearly independent with the vectors of B_1. Adding these vectors to B_1 we obtain $k + l - m$ linearly independent vectors. But as (6.3) means that $k + l - m = n$, we see that (6.2) and (6.3) are equivalent.

Consider a diffeomorphism f of the manifold M; let K, L be submanifolds of M. If v is a tangent vector, $v \in T_{x_0}K$, then

$$Df(x_0)v \in T_{f(x_0)}f(K).$$

As f is a diffeomorphism, the linear map $Df(x_0)$ is nonsingular for each $x_0 \in M$. Therefore, n-linearly independent vectors belonging to $T_{x_0}K, T_{x_0}L$ are mapped by $Df(x_0)$ on n-linearly independent vectors belonging to $T_{f(x_0)}f(K), T_{f(x_0)}f(L)$. We see that the following statement is valid.

Lemma 6.1. *If submanifolds K, L are transversal at x_0, and f is a diffeomorphism $M \to M$, then $f(K), f(L)$ are transversal at $f(x_0)$.*

Corollary. *Consider the system of differential equations* (1.1) *on a manifold M. Let φ be the flow generated by* (1.1). *Suppose that K, L are submanifolds of M invariant with respect to φ. If K, L are transversal at x_0, then K, L are transversal at any point $\varphi(t, x_0)$.*

Proof. For any $t \in \mathbb{R}$ the map $x \mapsto \varphi(t, x)$ is a diffeomorphism (see Section 3 in Chapter 1). □

2. We describe now some simple relations between transversality and hyperbolicity.

Consider a diffeomorphism $f : \mathbb{R} \to \mathbb{R}$. Let Δ be the diagonal of $\mathbb{R}\times\mathbb{R}$, and let grf be the map $\mathbb{R} \to \mathbb{R}\times\mathbb{R}$ defined by $grf(x) = (x, f(x))$. Evidently p is a fixed point of f if and only if $grf(p) \in \Delta$.

Lemma 6.2. *If p is a hyperbolic fixed point of f then the map grf is transversal to Δ at p.*

Proof. If p is a hyperbolic fixed point of f then $|f'(p)| \neq 1$. The linear map $Dgrf(p)$ maps $\mathbb{R}$ onto a line parallel to the vector $(1, f'(p))$. The space $T_{(p,p)}\Delta$ contains the vector $(1, 1)$. The vectors $(1, f'(p))$ and $(1, 1)$ are linearly independent, which implies the transversality of grf to Δ at p. □

Remark. Note that the transversality of grf to Δ at p does not imply the hyperbolicity of p. A simple example: a fixed point $x = 0$ of the diffeomorphism $f(x) = -x$.

Consider two sets of diffeomorphisms $f : \mathbb{R} \to \mathbb{R}$:
$F_1 = \{f$: every fixed point is hyperbolic $\}$,
$F_2 = \{f$: for any $k \geq 0$, $gr(f^k) \pitchfork \Delta\}$.

Lemma 6.3. $F_1 = F_2$

Proof. It follows from Lemma 6.1 that $F_1 \subset F_2$. To show that $F_2 \subset F_1$ consider a diffeomorphism $f \in F_2$ and a periodic point p of f of period m. As $gr(f^m) \pitchfork \Delta$ we have that $(f^m)'(p) \neq 1$. If $(f^m)'(p) = -1$, then

$$(f^{2m})'(p) = (f^m)'(f^m(p))(f^m)'(p) = [(f^m)'(p)]^2 = 1.$$

That means that $gr(f^{2m})$ is not transversal to Δ at p. The contradiction we obtained shows that $F_2 \subset F_1$. □

For multidimensional diffeomorphisms the transversality of graphs to the diagonal is connected with a weaker property than hyperbolicity. Consider a diffeomorphism $f : \mathbb{R}^n \to \mathbb{R}^n$. Let Δ be the diagonal of $\mathbb{R}^n \times \mathbb{R}^n$; define grf by $grf(x) = (x, f(x))$.

We say that the fixed point p of f is simple if the matrix $Df(p) - E$ is nonsingular.

Lemma 6.4. *A fixed point p is simple if and only if grf is transversal to Δ at p.*

Proof. The transversality of grf to Δ at p means that

$$Dgrf(p)(\mathbb{R}^n) + T_{(p,p)}\Delta = \mathbb{R}^n \times \mathbb{R}^n \tag{6.4}$$

(we identify $T_x\mathbb{R}^m$ and $\mathbb{R}^m$). For $\xi \in \mathbb{R}^n$ we have

$$Dgrf(p)\xi = (\xi, Df(p)\xi).$$

So (6.4) is equivalent to the following statement: for any $(x, y) \in \mathbb{R}^n \times \mathbb{R}^n$ there exist $\xi, \alpha \in \mathbb{R}^n$ such that

$$(\xi, Df(p)\xi) + (\alpha, \alpha) = (x, y)$$

or, what is the same

$$x = \alpha + \xi, \ y = \alpha + Df(p)\xi. \tag{6.5}$$

It follows from (6.5) that

$$\alpha = x - \xi, \ y - x = (Df(p) - E)\xi. \tag{6.6}$$

If the matrix $Df(p) - E$ is nonsingular, we can solve (6.6) with respect to ξ, α for any x, y, so we obtain that (6.4) is valid. On the other hand, if grf is transversal to Δ at p, then (6.6) is solvable with respect to ξ, α for any x, y. Therefore, the matrix $Df(p) - E$ is nonsingular. □

3. Consider the autonomous system of differential equations (1.1) on a smooth closed n-dimensional manifold M or in a domain $G \subset \mathbb{R}^n$ having the properties described in Section 1 of Chapter 3. Suppose that all rest points and all closed trajectories of (1.1) are hyperbolic. Denote by P the set of rest points and closed trajectories of (1.1).

We say that the system (1.1) satisfies the transversality condition if for any $p, q \in P$ the manifolds $W^s(p), W^u(q)$ are transversal.

The systems for which all the trajectories of the set P are hyperbolic and the transversality condition holds, are called Kupka–Smale systems. We study the set of Kupka–Smale systems in Chapter 7.

Let p, q be two trajectories of the set P for a Kupka–Smale system. We write $p \to q$ if there is a point $x \in W^u(p) \cap W^s(q)$ such that $x \notin p \cup q$.

Lemma 6.5. *Let system* (1.1) *be a Kupka–Smale system. Suppose that for* $p, q \in P$ *we have* $p \to q$. *Then*

$$dim W^u(p) \geq dim W^u(q) + 1 \tag{6.7}$$

if q *is a rest point, and*

$$dim\, W^u(p) \geq dim\, W^u(q) \tag{6.8}$$

if q *is a closed trajectory.*

Proof. Consider a point $x \in W^u(p) \cap W^s(q)$, $x \notin p \cup q$. Evidently $F(x) \neq 0$. As the manifolds $W^u(p), W^s(q)$ are invariant, the trajectory $\varphi(t, x)$ belongs to each of them. Therefore, the tangent vector of $\varphi(t, x)$ at $t = 0, F(x)$, belongs to $T_x W^u(p)$ and to $T_x W^s(q)$. We then have

$$\dim(T_x W^u(p) \cap T_x W^s(q)) \geq 1. \tag{6.9}$$

Use the condition of transversality of $W^u(p)$ and $W^s(q)$ at x in the form (6.3):

$$\dim W^u(p) + \dim W^s(q) - \dim(T_x W^u(p) \cap T_x W^s(q)) = n.$$

It follows from (6.9) that

$$\dim W^u(p) + \dim W^s(q) \geq n + 1 \tag{6.10}$$

or

$$\dim W^u(p) \geq n + 1 - \dim W^s(q). \tag{6.11}$$

We also know that for a hyperbolic rest point q, dim $W^u(q) = n-$ dim $W^s(q)$ (see (4.49)), and that for a hyperbolic closed trajectory q, dim $W^u(q) = n+1-$ dim $W^s(q)$ (see (5.6)). So, (6.11) implies (6.7) and (6.8). □

Corollary. *If q is a rest point and $\dim W^u(p) = \dim W^u(q)$ then $p \to q$ is impossible (in particular, $q \to q$ is impossible).*

Remark. For a closed hyperbolic trajectory p there can exist a point x such that $x \notin p$, and x is a point of transversal intersection of $W^u(p)$ and of $W^s(p)$. We say in this case that x is a transversal homoclinic point of the trajectory p. The structure of trajectories in a neighborhood of a homoclinic trajectory is very complicated. In particular we show in Chapter 10 that any neighborhood of a transversal homoclinic trajectory contains an infinite set of different closed trajectories.

4. Consider now a diffeomorphism f either on a smooth closed n-dimensional manifold M or in a domain $G \subset \mathbf{R}^n$. Denote by P the set of periodic points of f, and suppose that all periodic points of f are hyperbolic. We say that the diffeomorphism f satisfies the transversality condition if for any $p, q \in P$ the manifolds $W^u(p), W^s(q)$ are transversal.

Note that in the case of diffeomorphisms there can exist a point x of transversal intersection of $W^u(p), W^s(q)$ for $p, q \in P$ such that

$$\dim(T_x W^u(p) \cap T_x W^s(q)) = 0. \tag{6.12}$$

Therefore, if we write the condition of transversality

$$\dim W^u(p) + \dim W^s(q) = n + \dim(T_x W^u(p) \cap T_x W^s(q))$$

we obtain the following inequality

$$\dim W^u(p) + \dim W^s(q) \geq n. \tag{6.13}$$

As dim $W^u(q) = n-$ dim $W^s(q)$, it follows from (6.12) that the existence of a point of transversal intersection of $W^u(p), W^s(q)$ implies

$$\dim W^u(p) \geq \dim W^u(q). \tag{6.14}$$

5. We are going to prove now a very important fact—the so-called λ-Lemma [19] (for cascades—Theorem 6.1 below, for flows—the corollary of Theorem 6.1).

Let $0 \in \mathbb{R}^n$ be a hyperbolic fixed point of a diffeomorphism f of class C^1. Suppose that $\mathbb{R}^n = \mathbb{R}^{n_1} \times \mathbb{R}^{n_2}$ with coordinates y in $\mathbb{R}^{n_1}$, z in $\mathbb{R}^{n_2}$ so that according to the decomposition $x = (y, z)$ the matrix $Df(0)$ is block diagonal:

$$Df(0) = \begin{pmatrix} A & 0 \\ 0 & B \end{pmatrix}.$$

Here A is $n_1 \times n_1$ matrix with eigenvalues $\lambda_1, \ldots, \lambda_{n_1}$, and $|\lambda_j| < 1$, $j = 1, \ldots, n_1$; B is a $n_2 \times n_2$ matrix with eigenvalues $\lambda_{n_1+1}, \ldots, \lambda_n$ and $|\lambda_j| > 1$, $j = n_1 + 1, \ldots, n$.

It follows from the Canonical Form Theorem that we can choose coordinates y, z so that the following inequalities hold:

$$a_0 = \|A\| < 1, \ \frac{1}{b_0} = \|B^{-1}\| < 1. \tag{6.15}$$

In a neighborhood U_0 of the origin we have

$$f(y, z) = (Ay + f_1(y, z), Bz + f_2(y, z))$$

where

$$f_i(0, 0) = 0, \ \frac{\partial f_i}{\partial (y, z)}(0, 0) = 0, \quad i = 1, 2.$$

Suppose that in U_0 the disc $W^s_{loc}(0)$ coincides with the subspace $\{z = 0\}$, and the disc $W^u_{loc}(0)$ coincides with the subspace $\{y = 0\}$. Then for small $|z|$ we have

$$f(0, z) = (0, Bz + f_2(0, z)),$$

hence $f_1(0, z) = 0$. Similarly for small $|y|$ $f_2(y, 0) = 0$. Consequently, in a neighborhood of the origin

$$\frac{\partial f_2}{\partial y}(y, 0) = 0, \ \frac{\partial f_1}{\partial z}(0, z) = 0. \tag{6.16}$$

We suppose that (6.16) is valid in U_0.

Theorem 6.1. *Let N be a smooth disc having a point of transversal intersection with $W^s(0)$. There exists $\Delta > 0$ having the following property. Given $\varepsilon > 0$, there exists $m(\varepsilon)$ such that for $m \geq m(\varepsilon)$ we can find an embedding β_m of the disc*

$$Q = \{(y,z) : y = 0, |z| \leq \Delta\}$$

in $\mathbb{R}^n$ such that $\beta_m(Q) \subset f^m(N)$ and $\rho_1(\beta_m, id) < \varepsilon$ in $E^1(Q, \mathbb{R}^n)$.

Proof. Take $k > 0$ such that the following inequalities hold:

$$a = a_0 + k < 1,\ b = b_0 - k > 1,\ k < \frac{(b-1)^2}{8}. \tag{6.17}$$

Let V be a neighborhood of the origin such that for $(y,z) \in \overline{V}$ we have

$$\|\frac{\partial f_i}{\partial(y,z)}(y,z)\| < k,\ \ i = 1,2. \tag{6.18}$$

Consider a point q_0 of transversal intersection of N and $W^s(0)$. As $f^m(q_0) \underset{m\to\infty}{\longrightarrow} 0$, there exists a point $q = f^{m_0}(q_0) \in V$ of transversal intersection of $f^{m_0}(N)$ and $W^s(0)$. Choose a smooth n_2-dimensional disc $\widetilde{N} \subset f^{m_0}(N)$ such that $q \in \widetilde{N}$ and $\widetilde{N}$ is transversal to $W^s(0)$ at q (it is evidently possible). We are going to prove the statement of the theorem for $f^m(\widetilde{N})$ instead of $f^{m+m_0}(N)$. For simplicity we write N instead of $\widetilde{N}$.

Let v_0 be a unit tangent vector in T_qN. According to the decomposition $x = (y,z)$ we write $v_0 = (v_0^s, v_0^u)$. As N is transversal to $\{z = 0\}$ and as $\dim N = n_2$ we have that $v_0^u \neq 0$. Define the "inclination" of v_0 as

$$\lambda_0 = \frac{|v_0^s|}{|v_0^u|}.$$

The unit sphere of T_qN is compact; therefore there exists $\Lambda > 0$ such that for $v_0 \in T_qN$, $|v_0| = 1$, we have $\lambda_0 \leq \Lambda$.

Consider

$$Df(q) = \begin{pmatrix} A + \frac{\partial f_1}{\partial y} & \frac{\partial f_1}{\partial z} \\ \frac{\partial f_2}{\partial y} & B + \frac{\partial f_2}{\partial z} \end{pmatrix}.$$

As $q \in V \cap \{z = 0\}$ we have $\frac{\partial f_2}{\partial y}(q) = 0$. Thus, if $v_1 = Df(q)v = (v_1^s, v_1^u)$, we have

$$v_1^s = \left(A + \frac{\partial f_1}{\partial y}\right) v_0^s + \frac{\partial f_1}{\partial z} v_0^u, v_1^u = \left(B + \frac{\partial f_2}{\partial z}\right) v_0^u. \tag{6.19}$$

It follows from (6.15) and (6.19) that

$$|v_1^s| \leq (a_0 + k)|v_0^s| + k|v_0^u| \leq a|v_0^s| + k|v_0^u|. \tag{6.20}$$

Apply B^{-1} to the second equality in (6.19):

$$v_0^u + B^{-1}\frac{\partial f_2}{\partial z} v_0^u = B^{-1} v_1^u,$$

hence

$$\frac{1}{b_0}|v_1^u| \geq \|B^{-1}\| \cdot |v_1^u| \geq |v_0^u| - |B^{-1}\frac{\partial f_2}{\partial z} v_0^u| \geq |v_0^u| - \frac{k}{b_0}|v_0^u|,$$
$$|v_1^u| \geq (b_0 - k)|v_0^u| = b|v_0^u|. \tag{6.21}$$

To estimate the inclination of v_1 we use (6.20) and (6.21):

$$\lambda_1 = \frac{|v_1^s|}{|v_1^u|} \leq \frac{a|v_0^s| + k|v_0^u|}{b|v_0^u|} \leq \frac{a}{b}\lambda_0 + \frac{k}{b} \tag{6.22}$$

Consider $v_m = (v_m^s, v_m^u) = Df(f^{m-1}(q))v_{m-1}$,

$$\lambda_m = \frac{|v_m^s|}{|v_m^u|}.$$

Similar to (6.22), we obtain

$$\lambda_2 \leq \frac{a}{b}\lambda_1 + \frac{k}{b} \leq \left(\frac{a}{b}\right)^2 \lambda_0 + \frac{k}{b^2} + \frac{k}{b}, \ldots,$$
$$\lambda_m \leq \left(\frac{a}{b}\right)^m \lambda_0 + \frac{k}{b^m} + \ldots + \frac{k}{b}$$
$$\leq \left(\frac{a}{b}\right)^m \lambda_0 + \frac{k}{b-1} \tag{6.23}$$

Take m_1, such that the inequality

$$\left(\frac{a}{b}\right)^{m_1} \Lambda < \frac{b-1}{8}$$

holds. It follows from (6.17) and (6.23) that for any unit vector $v \in T_{f^m(q)}N$ with $m \geq m_1$ the inclination

$$\lambda_m \leq \frac{b-1}{4}. \tag{6.24}$$

By the continuity of the tangent plane of $T_x(f^{m_1}(N))$ with respect to x there exists a smooth disc $N_1 \subset f^{m_1}(N)$ such that $f^{m_1}(q)$ is a point of transversal intersection of N_1 and $W^s(0)$, and for any unit vector $v \in T_xN_1, x \in N_1$, the inclination λ of v satisfies the following inequality:

$$\lambda \leq \frac{b-1}{2} \tag{6.25}$$

(note that we use (6.24)).

Consider a disc $Q = \{(0, z) : |z| \leq \Delta\}$ belonging to V. Take any $\varepsilon_0 > 0$. Let

$$b_1 = \frac{b+1}{2},$$

and take ε_0 so small that $b_2 = \frac{b_1}{\sqrt{1+\varepsilon_0^2}} > 1$. It follows from (6.16) that $\frac{\partial f_1}{\partial z}(r) = 0$ for $r \in Q$. Consider such a neighborhood $V_0 \subset V$ of Q that

$$\left\|\frac{\partial f_1}{\partial z}(y, z)\right\| < k_1 = \frac{\varepsilon_0(b_1 - 1)}{2} \tag{6.26}$$

for $(y, z) \in \overline{V}_0$. Suppose that the disc N_1 considered above belongs to V_0. Consider $x \in N_1, v \in T_xN_1$. Let $v = (v^s, v^u)$, and let λ be the inclination of v. For the vector $v_1 = (v_1^s, v_1^u) = Df(x)v$ we have

$$v_1^s = \left(A + \frac{\partial f_1}{\partial y}\right) v^s + \frac{\partial f_1}{\partial z} v^u,$$

$$v_1^u = \frac{\partial f_2}{\partial y} v^s + \left(B + \frac{\partial f_2}{\partial z}\right) v^u$$

It follows from (6.17), (6.18), (6.26) that

$$|v_1^s| \leq a|v^s| + k_1|v^u|,$$

$$\frac{1}{b_0}|v_1^u| \geq |B^{-1}v_1^u| \geq (1 - \frac{k}{b_0})|v^u| - \frac{k}{b_0}|v^s|,$$

$$|v_1^u| \geq b|v^u| - k|v^s|,$$

$$\lambda_1 \leq \frac{a|v^s| + k_1|v^u|}{b|v^u| - k|v^s|} = \frac{a\lambda + k_1}{b - k\lambda}. \tag{6.27}$$

As $0 < k < 1$, it follows from (6.25) that

$$b - k\lambda > b - \frac{b-1}{2} = b_1. \tag{6.28}$$

Hence,

$$\lambda_1 \le \frac{a}{b_1}\lambda + \frac{k_1}{b_1}.$$

Similarly, if for $x \in N$ we have $f(x), \ldots, f^{m-1}(x) \in V$ and if we denote $v_2 = Df(f(x))v_1, \ldots, v_m = Df(f^{m-1}(x))v_{m-1}$, then

$$\lambda_m \le \left(\frac{a}{b_1}\right)^m \lambda + \frac{k_1}{b_1^m} + \ldots + \frac{k_1}{b_1} \le \left(\frac{a}{b_1}\right)^m_{\lambda} + \frac{k_1}{b_1 - 1}. \tag{6.29}$$

We obtain from (6.26) and (6.29) that there is an m_2 such that for $m \ge m_2$ we have $\lambda_m < \varepsilon_0$. If $m \ge m_2, f^m(x) \in V_0, v_{m+1} = Df(f^m(x))v_m$, then

$$\begin{aligned} \frac{|v_{m+1}|}{|v_m|} &= \frac{\sqrt{(v^s_{m+1})^2 + (v^u_{m+1})^2}}{\sqrt{(v^s_m)^2 + (v^u_m)^2}} \\ &= \frac{|v^u_{m+1}|}{|v^u_m|} \cdot \frac{\sqrt{1 + \lambda^2_{m+1}}}{\sqrt{1 + \lambda^2_m}}. \end{aligned} \tag{6.30}$$

It follows from (6.27) that $|v^u_{m+1}| \ge b_1 |v^u_m|$. Evidently,

$$\frac{\sqrt{1 + \lambda^2_{m+1}}}{\sqrt{1 + \lambda^2_m}} \ge \frac{1}{\sqrt{1 + \lambda^2_m}} \ge \frac{1}{\sqrt{1 + \varepsilon_0^2}}.$$

Hence, it follows from (6.30) that $|v_{m+1}| \ge b_2 |v_m|$. Therefore, for $m \ge m_2$ the map f uniformly expands the part of the disc $f^m(N_1)$ belonging to V_0. The tangent spaces of $f^m(N_1)$ at points $x \in V_0$ have a uniformly small slope to $W^u(0)$ (it is bounded by ε_0). So for large m we can define a map

$$\beta_m : Q \to f^m(N_1)$$

generated by the canonical projection onto Q. It follows from the description of the structure of a neighborhood of a hyperbolic fixed point given after Theorem 5.1 that f uniformly contracts vectors parallel to

$W^s(0)$ (in proper coordinates). Hence, there exists m_3 such that for $m \geq m_3$

$$|x - \beta_m(x)| < \varepsilon_0$$

for $x \in Q$. We see that for large m the map β_m has all the properties described in the statement of the theorem. □

Corollary [λ- Lemma for a hyperbolic rest point of a flow]. *Let $0 \in \mathbf{R}^n$ be a hyperbolic rest point of the system* (1.1), *and F be of class C^1. Suppose that $\mathbf{R}^n = \mathbf{R}^{n_1} \times \mathbf{R}^{n_2}$ with coordinates y in $\mathbf{R}^{n_1}$, z in $\mathbf{R}^{n_2}$ so that for a neighborhood U_0 of the origin*

$$W^s_{loc}(0) = \{z = 0\} \cap U_0, W^u_{loc}(0) = \{y = 0\} \cap U_0.$$

Suppose also that N is a smooth disc having a point of transversal intersection with $W^s(0)$. For $T > 0$, let

$$N_T = \bigcup_{t \geq T,\ x \in N} \varphi(t, x). \tag{6.31}$$

Then there is a $\Delta > 0$ having the following property. Given $\varepsilon > 0$ there exists $\tau(\varepsilon)$ such that for $T \geq \tau(\varepsilon)$ we can find an embedding β_T of the disc

$$Q = \{(y, z) : y = 0, |z| \leq \Delta\}$$

in $\mathbf{R}^n$ such that $B_T(Q) \subset N_T$, and $\rho_1(\beta_T, id) < \varepsilon$ in $E^1(Q, \mathbf{R}^n)$.

Proof. It follows from our assumptions that in a neighborhood of the origin we can write the system (1.1) as

$$\begin{aligned} \dot{y} &= A_1 y + F_1(y, z), \\ \dot{z} &= A_2 z + F_2(y, z). \end{aligned} \tag{6.32}$$

Here A_1 is a $n_1 \times n_1$ matrix such that the eigenvalues λ_j of A_1 satisfy $Re\lambda_j < 0$, A_2 is a $n_2 \times n_2$ matrix such that the eigenvalues λ_j of A_2 satisfy $Re\lambda_j > 0$. The functions F_1, F_2 satisfy (4.3). According to the decomposition $x = (y, z)$ we write $\varphi = (\varphi_1, \varphi_2)$ for the flow φ of (6.32).

We claim that the trajectory of (6.32) with initial conditions $y(0) = \tilde{y}, z(0) = \tilde{z}$ is given by

$$\begin{aligned} \varphi_1(t,\tilde{y},\tilde{z}) &= e^{A_1 t}\tilde{y} + Y(t,\tilde{y},\tilde{z}), \\ \varphi_2(t,\tilde{y},\tilde{z}) &= e^{A_2 t}\tilde{z} + Z(t,\tilde{y},\tilde{z}), \end{aligned} \tag{6.33}$$

where

$$\begin{aligned} Y(t,0,0) = 0,\ Z(t,0,0) = 0, \\ \frac{\partial Y}{\partial(\tilde{y},\tilde{z})}(t,0,0) = 0,\ \frac{\partial z}{\partial(\tilde{y},\tilde{z})}(t,0,0) = 0. \end{aligned} \tag{6.34}$$

Let us substitute φ_1, φ_2 in (6.32). Consider the first of the equalities (we omit arguments):

$$\dot{\varphi}_1 = A_1\varphi_1 + F_1(\varphi_1, \varphi_2) \tag{6.35}$$

We consider (6.35) as a nonhomogenous linear system of differential equations such that the corresponding homogenous system is

$$\dot{y} = A_1 y.$$

Then we have

$$\varphi_1 = e^{A_1 t}\tilde{y} + \int_0^t e^{A_1(t-s)} F_1(\varphi_1, \varphi_2) ds. \tag{6.36}$$

Denote by $Y(t,\tilde{y},\tilde{z})$ the second term in (6.36). Then, as $\varphi_1(t,0,0) = 0$, $\varphi_2(t,0,0) = 0$, we see that $Y(t,0,0) = 0$.

As

$$\frac{\partial Y}{\partial \tilde{y}} = \int_0^t e^{A_1(t-s)} \left[\frac{\partial F_1}{\partial y} \cdot \frac{\partial \varphi_1}{\partial \tilde{y}} + \frac{\partial F_1}{\partial z} \cdot \frac{\partial \varphi_2}{\partial \tilde{y}} \right] ds$$

it follows that

$$\frac{\partial Y}{\partial \tilde{y}}(t,0,0) = 0.$$

Similarly we define $Z(t,\tilde{y},\tilde{z})$ and prove all the equalities in (6.34).

As

$$\|e^{A_1 t}\| \underset{t\to+\infty}{\longrightarrow} 0,\ \|e^{-A_2 t}\| \underset{t\to+\infty}{\longrightarrow} 0$$

there exists $\theta > 0$ such that

$$\|e^{A_1\theta}\| < 1, \|e^{-A_2\theta}\| < 1.$$

Therefore, the diffeomorphism $f(x) = \varphi(\theta, x)$ satisfies the assumptions of Theorem 6.1. It is easy to see now that the λ-Lemma for flows is a consequence of Theorem 6.1. □

We leave it to the reader to formulate an analogous statement for the case of a hyperbolic closed trajectory of a flow.

Remark. It was shown in Section 2 of Chapter 4 that if p is a hyperbolic rest point of an autonomous system of differential equations of class C^r, $r \geq 2$, then there exist coordinates in a neigborhood of p defined by a diffeomorphism h of class C^r and such that:

(1) $h(p) = 0, W^s_{loc}(0), W^u_{loc}(0)$ are discs in coordinate subspaces;

2) in the coordinates the corresponding system of differential equations is of class C^{r-1}.

So the given proof of the λ-Lemma is valid for systems of class C^2. Indeed the λ-Lemma is true for hyperbolic rest points and for hyperbolic closed trajectories of systems of class C^1. We shall use the result for systems of class C^1.

Theorems 6.2 and 6.4 below are important consequences of the λ-Lemma.

Theorem 6.2. *Let p be a hyperbolic rest point or a hyperbolic closed trajectory of an autonomous system of differential equations of class C^1. Suppose that N is a smooth disc having a point of transversal intersection with $W^s(p)$. Then for any $T > 0$*

$$W^u(p) \subset \overline{N_T}$$

(the set N_T is defined by (6.31)).

Proof. Let us suppose that p is a rest point. Take $T > 0$ and a point $x_0 \in W^u(p)$. Consider the disc $Q \subset W^u(p)$ described in the statement of the λ-Lemma. There exists $\tau < 0$ such that $\varphi(\tau, x_0) \in$ Int Q (we take Int Q with respect to the inner topology of $W^u(p)$).

Take a sequence $t_m \to +\infty$. It follows from the λ-Lemma that any neighborhood of $\varphi(\tau, x_0)$ contains a point from N_{t_m} for large m—take, for example, the point $\beta_{t_m}(\varphi(\tau, x_0)) = \xi_m$. Evidently $\xi_m \to \varphi(\tau, x_0)$ as $m \to \infty$. By the continuity of φ we have $\varphi(-\tau, \xi_m) \to x_0$ as $m \to \infty$. It follows from (6.31) that $\varphi(-\tau, \xi_m) \in N_{t_m - \tau}$. For large m we have $t_m - \tau > T$ and $N_{t_m - \tau} \subset N_T$. □

6. Let h be a smooth embedding of a closed ball $B = \{|\xi| \le a\} \subset \mathbb{R}^k$ in a smooth manifold M. Consider the smooth closed disc $K = h(B)$. The definition of the tangent space $T_x K$ of K for $x = h(\xi)$, $|\xi| = a$, is similar to the definition of $T_x M$. The space $T_x K$ is a K-dimensional linear subspace of $T_x M$. So we can define transversality of two smooth closed discs K_1, K_2 in M repeating the definition of transversality of submanifolds of M.

We state the following important result without a proof (a proof can be found in [32]).

Theorem 6.3. *Let K_1, K_2 be two transversal smooth closed discs in a smooth manifold M. There exists $\delta > 0$ such that for any smooth embeddings $h_i \in E^1(K_i, M)$, $i = 1, 2$, with $\rho_1(h_i, id) < \delta$ the discs $h_1(K_1), h_2(K_2)$ are transversal. If Int $K_1 \cap$ Int $K_2 \ne \emptyset$ (we take interiors with respect to the inner topologies) then we can find $\delta > 0$ such that if $\rho_1(h_i, id) < \delta$ then $h_1(K_1) \cap h_2(K_2) \ne \emptyset$.*

Theorem 6.4. *Let system* (1.1) *be a Kupka–Smale system. If $p, q, r \in P$, and $p \to q, q \to r$, then $p \to r$.*

Proof. Suppose that q is a rest point. Then take smooth discs $N_p \subset W^u(p), N_r \subset W^s(r)$ such that N_p intersects $W^s(q)$ transversally, and N_r intersects $W^u(q)$ transversally. It follows from the λ-Lemma that there exists a disc $Q_p \subset W^u(q)$ such that for any $\varepsilon > 0$ there is an embedding $\beta_p^\varepsilon \in E^1(Q_p, \mathbb{R}^n)$ with $\rho_1(\beta_p^\varepsilon, id) < \varepsilon$, and with $\beta_p^\varepsilon(Q_p) \subset W^u(p)$ (we use here that $W^u(p)$ is invariant). Now change t for $-t$ in (1.1). This change of time transforms stable manifolds into unstable manifolds. Applying again the λ-Lemma we can find a disc $Q_r \subset W^s(q)$ such that for any $\varepsilon > 0$ there exists an embedding $\beta_r^\varepsilon \in E^1(Q_r, \mathbb{R}^n)$ with $\rho_1(\beta_r^\varepsilon, id) < \varepsilon$, and with $\beta_r^\varepsilon(Q_r) \subset W^s(r)$. The point q is a point of transversal intersection of the discs Q_p, Q_r. It follows now from Theorem 6.3 that if ε is small enough then there exists a point of transversal intersection of $\beta_r^\varepsilon(Q_r), \beta_p^\varepsilon(Q_p)$. Evidently this point belongs to $W^u(p) \cap W^s(r)$. □

Chapter 7

The Kupka–Smale Theorem

1. Consider an autonomous system of differential equations

$$\dot{x} = F(x). \tag{7.1}$$

Let $X = X^1(M)$ where M is a smooth closed manifold, or let $X = X^1_+(G)$, $G \subset \mathbb{R}^n$ (see Section 1 of Chapter 3).

We defined Kupka–Smale systems in Chapter 6; the system (7.1) is a Kupka–Smale system if its rest points and closed trajectories are hyperbolic and if the transversality condition holds. Denote KS as the set of Kupka–Smale systems.

In 1963, I. Kupka and S. Smale published independent proofs of the following statement ([11, 37]).

Theorem 7.1. [the Kupka–Smale Theorem]. *The set KS is residual in X.*

So a generic system of differential equations is a Kupka–Smale system. We show in Chapter 9 that if a system is structurally stable then it belongs to the set KS.

Below we give a sketch of a proof of Theorem 7.1. For simplicity we consider the case $X = X^1_+(G)$, $G \subset \mathbb{R}^n$.

Let p be a hyperbolic rest point of system (7.1). We constructed in Section 3, Chapter 4 a closed n_1-dimensional disc $\Sigma_0 \subset W^s_{loc}(p)$ such that any trajectory in $W^s(p)\backslash p$ has exactly one point of intersection with the set Σ, the boundary of Σ_0 (we consider the boundary with respect to the inner topology of $W^s(p)$). For $k \in \mathbb{Z}$, $k \geq 0$ define

$$W^s(p,k) = \sum_0 \cup \bigcup_{\sigma\in\Sigma, t\in(0,k]} \varphi(-t,\sigma).$$

The set $W^s(p,k)$ is compact and coincides with the image $b^s(\{|w| \leq k+1\})$, here b^s is the immersion $\mathbb{R}^{n_1} \to \mathbb{R}^n$ constructed in Section 3, Chapter 4. Evidently,

$$W^s(p) = \bigcup_k W^s(p,k).$$

Similarly, we define sets $W^u(p,k)$ as the images $b^u(\{|w| \leq k+1\})$ for the rest point p, and analogous sets $W^s(p,k), W^u(p,k)$ for a closed trajectory p of the system (7.1).

Denote P as the set of all rest points and closed trajectories of (7.1). By P_0 we denote the set of all rest points, and by P_k, $k > 0$, we denote the set of all rest points and of closed trajectories p such that the least period of p is less than or equal to k.

Define for $k \geq 0$ the following subset A_k of X; system (7.1) belongs to A_k if all the trajectories in P_k are hyperbolic, and if for any $p, q \in P_k$ the submanifolds with boundary $W^u(p,k), W^s(q,k)$ are transversal. It is easy to see that

$$KS = \bigcap_{k \geq 0} A_k.$$

So to prove Theorem 7.1 it is enough to show that any A_k is open and dense in X. We discuss the openness of A_k in Section 2, and the density of A_k in Section 3.

2. We divide the proof of the openness of A_k into several lemmas. Suppose that system (7.1) belongs to A_k.

Lemma 7.1. *The set P_k consists of a finite number of trajectories.*

Proof. Consider a rest point p of (7.1). The rest point p is hyperbolic; it follows from Theorem 4.4 that there exists a neighborhood U of p such that U contains no complete trajectories of (7.1) different from p. It follows from the continuity of the flow φ of (7.1) that there exists a neighborhood $U_k(p)$ of p such that $\varphi(t,x) \in U$ for $x \in U_k(p)$, $t \in [0,k]$. If a trajectory $q \in P_k, q \neq p$, intersects $U_k(p)$, then $q \subset U$, which contradicts Theorem 4.4. So the rest point p has a neighborhood $U_k(p)$ such that for any $q \in P_k, q \neq p$, we have $q \cap U_k(p) = \emptyset$.

If p is a hyperbolic closed trajectory of (7.1) belonging to the set P_k, we prove the existence of a neighborhood $U_k(p)$ with analogous properties using the Poincaré transformation of p.

The union of trajectories $p \in P_k$ is closed, so the existence of neighborhoods $U_k(p)$ implies the finiteness of P_k. □

Consider a system

$$\dot{x} = \widetilde{F}(x) \tag{7.2}$$

in X, and define for (7.2) the set $\widetilde{P}_k$ analogous to the set P_k for (7.1).

Lemma 7.2. *Let $P_k = \{p_1, \dots, p_m\}$. Given any neighborhood $U_1, \dots, U_m$ of $p_1, \dots p_m$ there exists a neighborhood W of (7.1) in X such that for a system (7.2)from W and for any $\tilde{p} \in \widetilde{P}_k$ we have*

$$\tilde{p} \subset U_1 \cup \dots \cup U_m.$$

Proof. Consider the compact $G' = \overline{G} \backslash (U_1 \cup \cdots U_m)$. System (7.1) has no rest points in G', so that there exists $\alpha > 0$ such that $|F(x)| \geq \alpha$ for $x \in G'$. Find a neighborhood W_0 of (7.1) in X and a number $M > 0$ such that

$$|\frac{\partial \widetilde{F}}{\partial x}(x)\widetilde{F}(x)| \leq M \tag{7.3}$$

for any system (7.2) from W_0 and any $x \in \overline{G}$. Define

$$\beta = \frac{\alpha}{2M}.$$

Consider the map $g : (0, k] \times G' \to \mathbb{R}$, defined by

$$g(t, x) = |\varphi(t, x) - x|.$$

There exists $\Delta > 0$ such that $g(t, x) \geq \Delta$ for $(t, x) \in [\beta, k] \times G'$. Denote by $\tilde{\varphi}$ the flow generated by system (7.2) and let

$$\tilde{g}(t, x) = |\tilde{\varphi}(t, x) - x|.$$

It follows from the basic course of differential equations that solutions are continuous with respect to parameters. So it is easy to see that there exists a neighborhood W_1 of (7.1) in X such that for any system (7.2) in W_1 we have

$$\tilde{g}(t,x) \geq \frac{\Delta}{2}$$

for $(t,x) \in [\beta,k] \times G'$. Find $W_2 \subset W_0$ such that for any system (7.2) in W_2 we have $|\widetilde{F}(x)| \geq \frac{\alpha}{2}$ for $x \in G'$ Differentiating

$$\frac{d}{dt}\tilde{\varphi}(t,x) = \widetilde{F}(\tilde{\varphi}(x,x))$$

with respect to t we obtain

$$\frac{d^2}{dt^2}\tilde{\varphi}(t,x) = \frac{\partial \widetilde{F}}{\partial x}(\tilde{\varphi}(t,x))\widetilde{F}(\tilde{\varphi}(t,x)). \tag{7.4}$$

Write the following Taylor formula for $\tilde{\varphi}$:

$$\tilde{\varphi}(t_0,x) = x + \widetilde{F}(x)t_0 + \frac{d^2}{dt^2}\tilde{\varphi}(t,x)\Big|_{t=\theta} \cdot \frac{t_0^2}{2}$$

where $\theta \in (0,t_0)$. We obtain from (7.3), (7.4)

$$|\tilde{\varphi}(t_0,x) - x - \widetilde{F}(x)t_0| \leq \frac{Mt_0^2}{2}.$$

Hence, for $x \in G'$

$$\tilde{g}(t_0,x) \geq \frac{\alpha t_0}{2} - \frac{Mt_0^2}{2} = \frac{t_0(\alpha - Mt_0)}{2}.$$

As $Mt_0 < 0,5\alpha$ for $t_0 \in (0,\beta)$, we obtain that $\tilde{g}(t,x) > 0$ for $(t,x) \in (0,\beta) \times G'$. Now take $W = W_1 \cap W_2$. □

Lemma 7.3. *Let p be a hyperbolic rest point of (7.1). There exists a neighborhood U of p and a neighborhood W of system (7.1) in X such that any system (7.2) in W has exactly one rest point $\tilde{p}$ in U, and $\tilde{p}$ is a hyperbolic rest point of (7.2).*

Proof. Consider for definiteness $p = 0$. In a neighborhood of the origin we can write

$$F(x) = \frac{\partial F}{\partial x}(0)x + g(x),$$

where

$$g(0) = 0, \ \frac{\partial g}{\partial x}(0) = 0. \tag{7.5}$$

It follows from the hyperbolicity of the rest point $x = 0$ that the eigenvalues of the matrix

$$\frac{\partial F}{\partial x}(0)$$

are nonzero so that this matrix is nonsingular. It is well-known that if a matrix A is nonsingular, and $\|A - B\|$ is small, then $\|A^{-1} - B^{-1}\|$ is small. So for vector-functions $\widetilde{F}, C^1$-close to F the matrices

$$\left(\frac{\partial \widetilde{F}}{\partial x}(0)\right)^{-1}$$

are close to the matrix

$$\left(\frac{\partial F}{\partial x}(0)\right)^{-1}$$

Therefore, there exists $N > 0$ and a neighborhood W_1 of (7.1) in X such that for any system (7.2) in W_1 we have

$$\left\| \left(\frac{\partial \widetilde{F}}{\partial x}(0)\right)^{-1} \right\| \leq N.$$

Take

$$\Delta = \frac{1}{N}.$$

Let us write

$$\widetilde{F}(x) = \widetilde{F}(0) + \frac{\partial \widetilde{F}}{\partial x}(0)x + \tilde{g}(x),$$

where

$$\tilde{g}(0) = 0, \frac{\partial \tilde{g}}{\partial x}(0) = 0.$$

It follows from (7.5) and from equality

$$\frac{\partial \tilde{g}(x)}{\partial x} = \frac{\partial F}{\partial x}(0) - \frac{\partial \widetilde{F}}{\partial x}(0) + \frac{\partial \widetilde{F}}{\partial x}(x) - \frac{\partial F}{\partial x}(x) + \frac{\partial g}{\partial x}(x)$$

that there exists a neighborhood U_1 of the origin and a neighborhood $W_2 \subset W_1$ of system (7.1) in X such that

$$\|\frac{\partial \tilde{g}}{\partial x}(x)\| < \Delta \tag{7.6}$$

for $x \in U_1$ and for $\widetilde{F} \in W_2$.

Choose $r > 0$ such that the closed ball

$$D_r = \{|x| \leq r\}$$

belongs to U_1. Find a neighborhood $W \subset W_2$ of the system (7.1) such that for $\widetilde{F} \in W$ the inequality

$$N|\widetilde{F}(0)| < \frac{r}{2}$$

holds and for $x \in D_r, \widetilde{F} \in W$ the eigenvalues of matrices

$$\frac{\partial \widetilde{F}}{\partial x}(x)$$

have nonzero real parts (this is possible, as the eigenvalues are continuous with respect to variations of elements of matrices). Take $\widetilde{F} \in W$ and define the map H by

$$H(x) = -\left(\frac{\partial \widetilde{F}}{\partial x}(0)\right)^{-1} (\widetilde{F}(0) + \tilde{g}(x)).$$

Evidently fixed points of H coincide with rest points of (7.2). As $\tilde{g}(0) = 0$, it follows from (7.6) that for $x \in D_r$

$$|\tilde{g}(x)| \leq \Delta r.$$

Then

$$|H(x)| \leq N|\widetilde{F}(0)| + N\Delta r \leq r$$

and consequently, H maps the ball D_r into itself.

As

$$|H(x_1) - H(x_2)| \leq N \max_{x \in Dr} \left\|\frac{\partial \tilde{g}}{\partial x}\right\| \cdot |x_1 - x_2| \leq \frac{|x_1 - x_2|}{2},$$

H is a contraction on D_r. Hence, there exists a unique fixed point $\tilde{p}$ of H in D_r. It follows from the choice of W that $\tilde{p}$ is a hyperbolic rest point of (7.2). □

Remark 1. The given proof of Lemma 7.3 is close to standard proofs of the Implicit Function Theorem. It was important for us to show that estimates in this proof are uniform with respect to systems (7.2) in a neighborhood of system (7.1) (this uniformity is not usually noted in basic courses of calculus).

2. An analogous but more complicated proof shows that the following statement is true.

If $p \in P_k$ is a hyperbolic closed trajectory of (7.1) then there exists a neighborhood U of p and a neighborhood W of (7.1) such that for any system (7.2) in W there exists not more than one trajectory $\tilde{p} \in \widetilde{P}_k$ with $\tilde{p}_k \cap U \neq \emptyset$, and if a trajectory $\tilde{p}$ having the described properties exists, then $\tilde{p}$ is hyperbolic.

The words "not more than one" are related to the following. If the least period of p equals to k, then there exists a closed trajectory $\tilde{p}$ of (7.2) near p (if the neigborhood W is small enough), but the least period of $\tilde{p}$ may be more than k, so $\tilde{p} \notin \widetilde{P}_k$. The following statement is a consequence of lemmas 7.1–7.3.

Lemma 7.4. *There exists a neighborhood W_0 of system (7.1) in X such that for any system (7.2) in W_0 every trajectory in $\widetilde{P}_k$ is hyperbolic.*

To finish the proof of the openness of A_k we need the following result on continuous dependence of compact subsets of stable and unstable manifolds with respect to C^1-small perturbations of the system.

Let p be a hyperbolic rest point or a hyperbolic closed trajectory of system (7.1). Consider an immersion $b^s : \Sigma \to \mathbb{R}^n$ such that $b^s(\Sigma) = W^s(p)$ (see Chapters 4 and 5); here $\Sigma = \mathbb{R}^m$ if p is a rest point, and Σ is a fibering over S^1 with fiber $\mathbb{R}^m$ if p is a closed trajectory.

Theorem 7.2. *Let D be a compact submanifold of the manifold Σ such that $b^s(D) = W^s(p, k)$. Given any $\varepsilon > 0$ and any neighborhood*

U of p there exists a neighborhood W of system (7.1) such that any system (7.2) in W has the following property. If p is a rest point of (7.1) *then there is a unique rest point $\tilde{p}$ of* (7.2) *in U; if p is a closed trajectory of* (7.1) *then there is a unique closed trajectory $\tilde{p}$ of* (7.2) *in U. Furthermore, there exists an embedding $\tilde{g}^s \in E^1(D, \mathbb{R}^n)$ such that $\tilde{g}^s(D) = W^s(p,k)$ and $\rho_1(g^s, \tilde{g}^s) < \varepsilon$.*

A proof of this theorem can be found in [8].

Remark. An analogous statement is true for $W^u(p,k)$.

The openness of the condition of transversality of intersections of $W^u(p,k), W^s(q,k)$ for $p, q \in P_k$ follows from Lemma 7.1 and from Theorems 6.3 and 7.2.

3. We give a sketch of a proof of the density of A_k in the case $k = 0$. This is the simplest case, but in the following proof we use the basic ideas that are needed to prove the density of A_k for $k > 0$.

Theorem 7.3, which we formulate below, is one of the main results of the theory of singularities (a proof can be found in [15]).

Consider a map $f : G \to \mathbb{R}^n$, $G \subset \mathbb{R}^n$, of class C^1. We say that $x_0 \in G$ is a critical point of f if

$$\det \frac{\partial f}{\partial x}(x_0) = 0. \tag{7.7}$$

Theorem 7.3 [The Sard–Brown Theorem]. *If S is the set of critical points of f, then mes $f(S) = 0$.*

Lemma 7.5. *If $x_0 \in G$ is a nonisolated rest point of system* (7.1), *then x_0 is a critical point of the map*

$$F : G \to \mathbb{R}^n.$$

Proof. The function F is of class C^1 so that we can write

$$F(x) = F(x_0) + \frac{\partial F}{\partial x}(x_0)(x - x_0) + g(x), \tag{7.8}$$

where

$$\frac{|g(x)|}{|x - x_0|} \underset{x \to x_0}{\longrightarrow} 0. \tag{7.9}$$

Consider a sequence x_m of rest points of (7.1) such that $x_m \to x_0$ as $m \to \infty$. It follows from (7.8) and (7.9) that

$$\frac{\partial F}{\partial x}(x_0)\frac{x_m - x_0}{|x_m - x_0|} \underset{m \to \infty}{\longrightarrow} 0. \tag{7.10}$$

Define vectors

$$\xi_m = \frac{x_m - x_0}{|x_m - x_0|}.$$

As the sphere $\{|x| = 1\}$ in $\mathbb{R}^n$ is compact, there is a limit point ξ of the sequence ξ_m. We obtain from (7.10) that

$$\frac{\partial F}{\partial x}(x_0)\xi = 0.$$

As $\xi \neq 0$, we conclude that

$$\det \frac{\partial F}{\partial x}(x_0) = 0.\square$$

Lemma 7.6. *Given any neighborhood W of system (7.1) in X there exists a system (7.2) in W such that the set of rest points of (7.2) in G is finite and for every rest point x_0 of (7.2) in G we have*

$$\det \frac{\partial \widetilde{F}}{\partial x}(x_0) \neq 0.$$

Proof. Consider the map

$$F : G \to \mathbb{R}^n,$$

where $F(x)$ is the vector field of system (7.1). Denote by S the set of critical points of the map.

Take arbitrary $\varepsilon > 0$. As mes $F(S) = 0$ (it follows from Theorem 7.3), and the measure of ε-neighborhood of the origin is positive, we can find $a \in \mathbb{R}^n$ such that $|a| < \varepsilon$ and $a \notin F(S)$. Consider system (7.2) with $\widetilde{F}(x) = F(x) - a$. If x_0 is a rest point of (7.2) then $\widetilde{F}(x_0) = 0$, so $F(x_0) = a$. As $a \notin F(S)$, we have

$$\det\frac{\partial \widetilde{F}}{\partial x}(x_0) = \det\frac{\partial F}{\partial x}(x_0) \neq 0.$$

It follows from Lemma 7.5 that x_0 is an isolated rest point of (7.2). The set of rest points of (7.2) is closed, and any rest point is isolated; therefore the set of rest points in G is finite. □

Lemma 7.7. *Let A be an $n \times n$ matrix. Given $\varepsilon > 0$ there exists a matrix B such that $\|A - B\| < \varepsilon$ and for the eigenvalues λ_j of B inequalities $Re\lambda_j \neq 0,\ j = 1, \dots, n$, hold.*

Proof. Denote the eigenvalues of A by $\mu_1, \dots, \mu_n$. It follows from the Canonical Form Theorem that there exists a nonsingular matrix S such that

$$J = S^{-1}AS = \begin{pmatrix} \mu_1 & & 0 \\ & \ddots & \\ \dots & & \mu_n \end{pmatrix}$$

(J is triangular). Take $\varepsilon_0 > 0$ such that $\varepsilon_0\|S\| \cdot \|S^{-1}\| < \varepsilon$. Find numbers $\tilde{\mu}_1, \dots, \tilde{\mu}_n$ so that $|\tilde{\mu}_j| < \varepsilon_0, j = 1, \dots, n$, and $Re(\mu_j + \tilde{\mu}_j) \neq 0$, and let

$$\widetilde{J} = J + \operatorname{diag}(\tilde{\mu}_1, \dots, \tilde{\mu}_n).$$

The matrix $\widetilde{J}$ is triangular, so its eigenvalues are $\lambda_j = \mu_j + \tilde{\mu}_j, j = 1, \dots, n$. Take now $B = S\widetilde{J}S^{-1}$. The eigenvalues of B are also $\lambda_1, \dots, \lambda_n$ and evidently $\|A - B\| < \varepsilon$. □

Lemma 7.8. *Given any neighborhood W of system* (7.1) *there is a system* (7.2) *in W such that every rest point of* (7.2) *is hyperbolic.*

Proof. Find in W a system (7.2) such that the statement of Lemma 7.6 is valid for this system. For simplicity we suppose that the statement of Lemma 7.6 is valid for system (7.1).

The set of rest points of (7.1) in G is finite. Denote the rest points by $p_1, \dots, p_m$. Take one of them: $x = p_i$, suppose for simplicity that

$p_i = 0$. Choose a neighborhood U of $x = 0$ such that U contains no rest points of (7.1) different from $x = 0$. We can write (7.1) in U as

$$\dot{x} = Ax + f(x), \tag{7.11}$$

where

$$\frac{|f(x)|}{|x|} \underset{|x|\to 0}{\longrightarrow} 0.$$

It follows from our choice of the system that

$$\det A = \det \frac{\partial F}{\partial x}(0) \neq 0.$$

Hence there exists $\delta > 0$ such that $|Ax| \geq \delta |x|$. Take a neighborhood $U_0 \subset U$ of the origin such that

$$|f(x)| \leq \frac{\delta}{4}|x| \tag{7.12}$$

for $x \in U_0$. We can write any perturbed system (7.2) as

$$\dot{x} = Ax + f(x) + \tilde{f}(x), \tag{7.13}$$

where $\tilde{f}(x) = \widetilde{F}(x) - F(x)$. Therefore there exists a neighborhood $W_0 \subset W$ of (7.1) such that any system (7.2) in W_0 has the following property:

$$\left\| \frac{\partial \tilde{f}}{\partial x}(x) \right\| \leq \frac{\delta}{4}.$$

Consequently, if $\widetilde{F}(0) = 0$, then for (7.2) in W_0 we have

$$|\tilde{f}(x)| \leq \frac{\delta}{4}|x|. \tag{7.14}$$

Take arbitrary $\varepsilon > 0$. It follows from Lemma 7.7 that there exists a matrix A_1 such that $\|A - A_1\| < \varepsilon$ and for the eigenvalues λ_j of A_1 we have $Re\lambda_j \neq 0$. Consider a smoothing function $g \in C^\infty(\mathbb{R})$ such that

$$g(t) = \begin{cases} 1, & t \leq 1, \\ 0, & t \geq 2, \end{cases}$$

$0 < g(t) < 1$ for $t \in (1,2)$; $|g'(t)| \leq 2$. Fix $\Delta > 0$ and define the function $F_\Delta(x)$ by

$$F_\Delta(x) = Ax + g(\frac{x^2}{\Delta})(A_1 - A)x + f(x).$$

Estimates similar to ones used in the proof of Lemma 4.5 show that

$$\max_{x \in \overline{G}} \left[|g(\frac{x^2}{\Delta})(A_1 - A)x| + \left\| \frac{\partial}{\partial x} \left(g(\frac{x^2}{\Delta})(A_1 - A)x \right) \right\| \right] \to 0$$

as $\varepsilon \to 0$ and $\Delta \to 0$. Hence, there exists $\varepsilon > 0$ and $\Delta > 0$ such that the system

$$\dot{x} = F_\Delta(x) \tag{7.15}$$

is in W_0. Choose $\Delta > 0$ so small that $F_\Delta(x) = F(x)$ for $x \notin U_0$. We have $F_\Delta(x) = A_1 x + f(x)$ for $x^2 < \Delta$, so $x = 0$ is a hyperbolic rest point of (7.15).

If we write (7.15) in the form (7.13) then (7.14) is valid. Therefore for $x \in U_0$ we obtain

$$|F_\Delta(x)| = |Ax + f(x) + \tilde{f}(x)| \geq \delta|x| - \frac{\delta|x|}{4} - \frac{\delta|x|}{4} = \frac{\delta|x|}{2}.$$

Hence, system (7.15) has no rest points in U_0 different from $x = 0$. We perturbed system (7.1) in a domain containing no rest points different from p_i. It is easy to see that if we perturb analogously system (7.1) in a neighborhood of every rest point we obtain a system in W having only hyperbolic rest points. □

To finish the proof of density of A_0 in X we need the following statement.

Lemma 7.9. *Suppose that every rest point of system* (7.1) *is hyperbolic. Then in any neighborhood W of* (7.1) *we can find a system* (7.2) *such that for any rest points p, q of* (7.2) *the submanifolds $W^s(p,0), W^u(q,0)$, are transversal.*

To construct a perturbation of a system of differential equations resulting in prescribed perturbations of compact parts of stable and unstable manifolds is a more complicated problem than to construct a perturbation transforming isolated rest points into hypebolic ones. We realize below, in the proof of Lemma 9.2, a sort of a desired perturbation of a system.

Consider a hyperbolic rest point p of (7.1), let dim $W^s(p) = n_1$. Consider the closed ball $D = \{|w| \leq 1\} \subset \mathbb{R}^{n_1}$, and an embedding $b^s \in E^1(D, \mathbb{R}^n)$ such that $b^s(D) = W^s(p, 0)$. Using arguments close to the proof of Lemma 9.2 below (but more complicated) one can prove the following statement.

Lemma 7.10. *Given a neighborhood W of system (7.1) in X there exists $\varepsilon > 0$ having the following property. For any embedding $\tilde{b} \in E^1(D, \mathbb{R}^n)$ such that $\tilde{b}(0) = p$, $\rho_1(b^s, \tilde{b}) < \varepsilon$ there exists a system (7.2) in W such that $\tilde{b}(D) = \widetilde{W}^s(p, 0)$ (here $\widetilde{W}^s(p)$ is the stable manifold of the rest point p of (7.2)).*

To use Lemma 7.10 for the proof of Lemma 7.9 we need the following variant of the Transversality Theorem (one can find a proof of this theorem in [32]).

Theorem 7.4. *Let D be a manifold with boundary, and let C be a submanifold with boundary of a smooth manifold M. Let $b \in E^1(D, M)$. Any neighborhood of b in $E^1(D, M)$ contains an embedding $\tilde{b}$ transversal to C.*

Proof (of Lemma 7.9). Let $P_0 = \{p_1, \dots, p_m\}$. Consider an arbitrary neighborhood W of system (7.1) in X. It follows from Theorem 7.4 and from Lemma 7.10 that there exists a system in W such that for this system $W^s(p_1, 0)$ is transversal to $W^u(p_1, 0), \dots, W^u(p_m, 0)$ (it is easy to see that we can find an embedding $\tilde{b}$ described in Theorem 7.4 so that $\tilde{b}(p_1) = p_1$). Then we apply Theorem 7.4 and Lemma 7.10 to $W^s(p_2, 0)$. It follows from Theorem 6.3 and from Theorem 7.2 that C^1-small perturbations of the system do not destroy the transversality of $W^s(p_1, 0), W^u(p_1, 0), \dots, W^u(P_m, 0)$, so we can find in W a system having the following property: $W^s(p_1, 0)$ and $W^s(p_2, 0)$ are transversal to $W^u(p_1, 0), \dots, W^u(p_m, 0)$. The set P_0 is finite, hence repeating the described process m times we can find in W a system such that

$W^s(p_i, 0)$ is transversal to $W^u(p_j, 0)$ for any $p_i, p_j \in P_0$.

4. The Kupka–Smale Theorem is valid for cascades generated by diffeomorphisms. In the case of cascades technical details of proofs are simpler than in the case of flows.

Consider $X = \mathrm{Diff}^1(M)$ or $X = \mathrm{Diff}^1_+(G)$ (see §3). We say that a diffeomorphism $f \in X$ is a Kupka–Smale diffeomorphism if periodic points of f are hyperbolic and the transversality condition is satisfied.

Theorem 7.5. *The set of Kupka–Smale diffeomorphisms is residual in* X.

Chapter 8

The Closing Lemma

1. Consider the autonomous system of differential equations (1.1). Let $X = X^1(M)$ or let $X = X^1_+(G)$. Denote by $\Omega(F)$ the nonwandering set of system (1.1).

C. Pugh proved the following two results [25, 26].

Theorem 8.1. [The Closing Lemma]. *Suppose that $x_0 \in \Omega(F)$, $F(x_0) \neq 0$. Given a neighborhood W of (1.1) in X there exists a system*

$$\dot{x} = \widetilde{F}(x) \tag{8.1}$$

in W such that the trajectory of x_0 with respect to the system (8.1) is closed.

Theorem 8.2. [The Density Theorem]. *The set of systems (1.1) for which rest points and closed trajectories are dense in $\Omega(F)$, is residual in X.*

These theorems are of great importance for the theory of structural stability. The proofs given in [25 and 26], and also proofs published later by Pugh and other mathematicians are very complicated.

Below we give a proof of a more simple result (Theorem 8.3) and explain some difficulties in the proofs of Theorem 8.1.

Theorem 8.3. [The Weakened C^0-Closing Lemma] *Let $X = X^0(M), x_0 \in \Omega(F), F(x_0) \neq 0$. Given a neighborhood U of x_0 and a neighborhood W of (1.1) in X there exists a system (8.1) in W having a closed trajectory that intersects U.*

Remark. The word "weakened" means the following: we construct a perturbed system (8.1) so that it has a closed trajectory not through x_0 but through a neighborhoof of x_0. It is easy to see that if a closed trajectory is close to x_0 then one can perturb the system so that the trajectory of x_0 in the perturbed system is closed.

2. Let us prove Theorem 8.3. If the trajectory of x_0 in the system (1.1) is closed we have nothing to prove, so we consider the case of nontrivial nonwandering point x_0.

Take a neighborhood U of x_0. As x_0 is not a rest point it follows from the Tubular Flow Theorem [3] that there exists a diffeomorphism H conjugating the flow of (1.1) on U and the flow of the system

$$\dot{\xi} = 1, \dot{\eta} = 0 \tag{8.2}$$

on a neighborhood V of the origin of $\mathbf{R}^n$ (one must reduce U if necessary). Here ξ, η are coordinates in $\mathbf{R}^n$, ξ is 1-dimensional, and η is $(n-1)$-dimensional. We can choose H so that $H(x_0) = (0,0)$.

Take $a, b > 0$ such that the set

$$V_0 = \{(\xi, \eta) : |\xi| \leq a, |\eta| \leq b\}$$

is a subset of V. The disc

$$\Sigma = \{(\xi, \eta) : \xi = 0, |\eta| \leq b\}$$

is transversal to trajectories of (8.2). Denote $U_0 = H^{-1}(V_0)$. As x_0 is nonwandering, given $T_0 > 0$ there exists $x \in U_0$ and $T \in \mathbf{R}$ such that $T > T_0$ and $\varphi(T, x) \in U_0$. Denote

$$\varphi([0,T], x) = \bigcup_{t \in [0,T]} \varphi(t, x).$$

Evidently the set

$$H(\varphi([0,T], x_0) \cap U_0)$$

is a finite set of segments in V_0 parallel to the line $\{\eta = 0\}$. Consider the set of points of intersections of these segments with Σ:

$$\Sigma \cap H(\varphi([0,T], x_0) \cap U_0). \tag{8.3}$$

This set is finite. It is easy to see that one can find two points $(o, \eta^{(1)})$, $(0, \eta^{(2)})$ in this set having to following property: the segment Q with ends at $(0, \eta^{(1)}), (0, \eta^{(2)})$ contains in its interior no points of the set (8.3).

Suppose for definiteness that the positive trajectory through $(0, \eta^{(2)})$ intersects $H^{-1}(\Sigma)$ at $(0, \eta^{(1)})$. That means the following: there exists $T' \in (0, T)$ such that

$$(0, \eta^{(1)}) = H(\varphi(T', H^{-1}(0, \eta^{(2)}))).$$

Take a neighborhood Y of the segment Q in the hyperplane $\{\xi = 0\}$ such that Y contains no points of the set (8.3) different from the ends of Q. This is possible as the set (8.3) is finite.

Consider two functions:

(1) $\beta \in C^\infty(\mathbb{R}^{n-1})$ such that

$$\beta(\eta) = 1 \quad \text{for} \quad \eta \in Q,$$

$$0 < \beta(\eta) < 1 \quad \text{for} \quad \eta \in Y \backslash Q,$$

$$\beta(\eta) = 0 \quad \text{for} \quad \eta \notin Y$$

(we identify $\{\xi = 0\}$ and $\mathbb{R}^{n-1}$);

(2) $\alpha \in C^\infty(\mathbb{R})$ such that

$$\alpha(\xi) = 1 \quad \text{for} \quad \xi \in [\frac{a}{3}, \frac{2a}{3}],$$

$$0 < \alpha(\xi) < 1 \quad \text{for} \quad \xi \in (0, \frac{a}{3}) \cup (\frac{2a}{3}, a),$$

$$\alpha(\xi) = 0 \quad \text{for} \quad \xi \notin [0, a]$$

(we identifty $\{\eta = 0\}$ and $\mathbb{R}$).

We discussed the existence of functions similar to α in the proof of Lemma 4.5. We leave it to the reader to construct β.

Let

$$A = \int_0^a \alpha(\tau) d\tau.$$

Evidently,

$$A > \frac{a}{3} \tag{8.4}$$

Consider the following system of differential equations in V_0:

$$\dot{\xi} = 1, \dot{\eta} = \frac{\eta^{(2)} - \eta^{(1)}}{A}\alpha(\xi)\beta(\eta). \tag{8.5}$$

It follows from properties of α, β that the system (8.5) is of class C^∞ and systems (8.2) and (8.5) coincide for $(\xi, \eta) \notin (0, a) \times Y$.

Consider the following functions:

$$\xi(t) = t, \eta(t) = \eta^{(1)} + \frac{\eta^{(2)} - \eta^{(1)}}{A}\int_0^t \alpha(\tau)d\tau. \tag{8.6}$$

It follows from the second equality in (8.6) that $(0, \eta(t)) \in Q$ for $t \in [0, a]$, hence $\beta(\eta(t)) \equiv 1$. Therefore, (8.6) is a solution of (8.5) for $t \in [0, a]$. As $\xi(0) = 0$, $\eta(0) = \eta^{(1)}, \xi(a) = a, \eta(a) = \eta^{(2)}$, the trajectory of the solution (8.6) joins the points $(0, \eta^{(1)})$ and $(a, \eta^{(2)})$ not leaving $[0, a] \times Y$. Change the system (1.1) in $H^{-1}(V_0)$ mapping the trajectories of (8.5) by H^{-1}, and leaving (1.1) unchanged outside of $H^{-1}(V_0)$. Let this perturbed system be (8.1). As (8.5) differs from (8.2) only for $(\xi, \eta) \in (0, a) \times Y \subset \operatorname{Int} V_0$, the system (8.1) is properly defined. Denote by $\tilde{\varphi}$ the flow of (8.1). It follows from the construction that

$$\tilde{\varphi}(t, H^{-1}((a, \eta^{(2)}))) = \varphi(t, H^{-1}((a, \eta^{(2)})))$$

for $t \in [0, T' - a]$. To see that we take into account that these segments of trajectories do not intersect the set $H^{-1}((0, a) \times Y)$, and outside of this set the systems (1.1) and (8.1) coincide. As

$$\varphi(T' - a, H^{-1}((a, \eta^{(2)}))) = H^{-1}((0, \eta^{(1)}))$$

and $\tilde{\varphi}(t, H^{-1}((0, \eta^{(1)})))$ joins $H^{-1}((0, \eta^{(1)}))$ and $H^{-1}((a, \eta^{(2)}))$ when $t \in [0, a]$, we obtain that the trajectory $\tilde{\varphi}(t, H^{-1}((0, \eta^{(1)})))$ is closed.

The C^0-distance between systems (8.2) and (8.5) is bounded above by

$$\Delta = \sup_{(\xi,\eta) \in (0,a) \times Y} \frac{|\eta^{(2)} - \eta^{(1)}|}{A}\alpha(\xi)\beta(\eta) \leq \frac{3|\eta^{(2)} - \eta^{(1)}|}{a}$$

(we take (8.4) into account). We can choose $x \in U_0$ and T_0 so that $|\eta^{(2)} - \eta^{(1)}|$ and Δ are arbitrarily small. Denote by $\Xi(\xi, \eta)$ the vector field of the system (8.5) and by $\psi(t,\xi,\eta)$ the trajectory of (ξ,η) is this system. Differentiating the equality

$$\tilde{\varphi}(t,x) = H^{-1}(\psi(t,H(x)))$$

with respect to t at $t=0$ we obtain

$$\widetilde{F}(x) = \frac{\partial H^{-1}}{\partial(\xi,\eta)}\Xi(H(x)). \tag{8.7}$$

The set V_0 is compact; hence there exists $N>0$ such that

$$\|\frac{\partial H^{-1}}{\partial(\xi,\eta)}\| \leq N, \quad (\xi,\eta)\in V_0.$$

Then it is easy to see that

$$\rho_0(F,\widetilde{F}) \leq N\Delta$$

As we can make Δ arbitrarily small, that completes the proof of Theorem 8.3. □

Remark. To prove Theorem 8.3 we estimated Δ using estimates $|\alpha|\leq 1$, $|\beta|\leq 1$. To estimate $\rho_1(F,\widetilde{F})$ we need estimates of

$$\Delta_1 = \sup \|\frac{\partial \beta}{\partial \eta}\|.$$

As $\beta(\eta)=1$ for $\eta\in Q$ and $\beta(\eta)=0$ for $\eta\notin Y$, it is easy to see that

$$\Delta_1 \geq \frac{1}{d} \tag{8.8}$$

where d is the distance between Q and the nearest point of the set (8.3) not belonging to Q. So the right hand of (8.8) depends on the choice of points $(0,\eta^{(1)})$ and $(0,\eta^{(2)})$ we are going to "join." One of the main difficulties in the proof of Pugh [25] is to choose a "fine" pair of points being joined by a segment of a trajectory of a perturbed system. See the paper [31] describing the main ideas of the C^1-closing.

Chapter 9

Necessary Conditions for Structural Stability

1. In this chapter we give a sketch of a proof of the following result mentioned in Chapter 7.

Theorem 9.1. *If an autonomous system of differential equations*

$$\dot{x} = F(x) \tag{9.1}$$

in $X^1(M)$ *or in* $X^1_+(G)$ *is structurally stable, then* (9.1) *is a Kupka–Smale system.*

This theorem states that if a system is structurally stable then its rest points and closed trajectories are hyperbolic, and the transversality condition holds.

We prove below that rest points of structurally stable systems are hyperbolic. We leave it to the reader to establish the hyperbolicity of closed trajectories. Also, we prove the necessity of the transversality condition under some additional assumptions stated below.

Consider $X = X^1_+(G)$, $G \in \mathbb{R}^n$.

Lemma 9.1. *If system* (9.1) *is structurally stable then its rest points are hyperbolic.*

Proof. Consider a neigborhood U of the system (9.1) given by the definition of structural stability. It follows from Theorem 7.1 that there exists in U a system

$$\dot{x} = \widetilde{F}(x) \tag{9.2}$$

being a Kupka–Smale system.

To get a contradiction suppose that the system (9.1) has a non-hyperbolic rest point. Assume that this rest point is the origin. We can write the system (9.1) in a neighborhood of the origin as

$$\dot{x} = Ax + f(x) \tag{9.3}$$

where

$$A = \frac{\partial F}{\partial x}(0),\ f(0) = 0,\ \frac{\partial f}{\partial x}(0) = 0.$$

Consider a function $\gamma \in C^\infty(\mathbb{R})$ such that $\gamma(z) = 0$ for $z \leq 1$; $\gamma(z) = 1$ for $z \geq 2$; $0 < \gamma(z) < 1$ for $z \in (1, 2)$; $|\gamma'(z)| \leq 2$. As in the proof of Lemma 4.5 one can show that there exists $\Delta > 0$ having the following property: the system

$$\dot{x} = F^*(x) \tag{9.4}$$

given by

$$\dot{x} = Ax + \gamma(\frac{x^2}{\Delta})f(x)$$

for $x^2 \leq 2\Delta$, and coinciding with (9.1) for $x^2 > 2\Delta$, is in the neighborhood U of (9.1) in X. As the pairs of systems (9.1) and (9.2), (9.1) and (9.4) are topologically equivalent in G, systems (9.2) and (9.4) are also topologically equivalent in G. Any rest point of system (9.2) is hyperbolic. It follows from Theorem 4.4 that any rest point p of (9.2) has a neighborhood $V(p)$ containing no complete trajectories of (9.2) different from p. It is easy to see that the topological equivalence of (9.2) and (9.4) implies the existence of similar neighborhoods for rest points of (9.4).

For $x^2 < \Delta$, system (9.4) can be written as

$$\dot{x} = Ax.$$

We supposed that the rest point $x = 0$ of (9.1) is not hyperbolic. Hence, the matrix A has either a zero eigenvalue, or a pair of purely imaginary eigenvalues.

In the first case in suitable coordinates

$$A = \begin{pmatrix} 0 & \dots \\ 0 & \dots \\ \dots & \dots \\ 0 & \dots \end{pmatrix}$$

hence the points $(a, 0, \dots, 0)$ with $|a| < \sqrt{\Delta}$ are rest points of (9.4).

In the second case in suitable coordinates

$$A = \begin{pmatrix} 0 & -\omega & \dots \\ \omega & 0 & \dots \\ 0 & 0 & \dots \\ & \dots & \dots \\ 0 & 0 & \dots \end{pmatrix}$$

for some $\omega > 0$. In this case in any neighborhood of the origin there are closed trajectories of (9.4) corresponding to solutions $x_1 = a\cos\omega t$, $x_2 = a\sin\omega t$, $x_3 = \ldots = x_n = 0$ with small a. The contradiction we obtained completes the proof. □

It was mentioned that we prove the necessity of the transversality condition under some additional conditions of system (9.1). We suppose below that $F \in C^2(G)$, $G \subset \mathbb{R}^3$. The additional smoothness (C^2 instead of C^1) is essential for the proof, and the choice of dimension $n = 3$ allows us to simplify considerations without loosing the main ideas.

To get a contradiction suppose that there exists a point z of non-transversal intersection of the stable manifold $W^s(q)$ and of the unstable manifold $W^u(p)$ for two trajectories p, q belonging to the set of rest points and of closed trajectories of system (9.1). Consider a neighborhood U of (9.1) given by the definition of structural stability. Find a Kupka–Smale system (9.2) in U. Let h be a topological equivalence of (9.1) and (9.2). Denote $\tilde{p} = h(p), \tilde{q} = h(q)$. Evidently $\tilde{z} = h(z)$ is a point of intersection of $W^u(\tilde{p})$ and $W^s(\tilde{q})$. Let us show that dim $W^u(p)$ = dim $W^s(q) = 2$. The homeomorphism h maps stable (unstable) manifolds onto stable (respectively, unstable) manifolds, so it is enough to show that

$$\dim W^u(\tilde{p}) = \dim W^s(\tilde{q}) = 2.$$

If $\tilde{q}$ is a rest point, it follows from Lemma 6.5 that

$$\dim W^u(\tilde{p}) \geq \dim W^s(\tilde{q}) + 1.$$

If we suppose that $\dim W^u(\tilde{p}) \neq 2$ then we have two possibilities: dim $W^u(\tilde{p}) = 3$ or dim $W^u(\tilde{p}) = 1$ (in this case dim $W^u(\tilde{q}) = 0$, dim $W^s(\tilde{q}) = 3$). In each of these cases the dimension of one of the manifolds $W^u(p), W^s(q)$ is equal to the dimension of the phase space so that its intersection with other manifolds is transversal. The equality dim $W^s(\tilde{q}) = 2$ is proved similarly. We leave it to the reader to consider the case of closed trajectory $\tilde{q}$.

It follows from the Tubular Flow Theorem that there exists a diffeomorphism H of class C^2 conjugating the flow φ on neighborhood V of z and the flow of system

$$\dot{u}_1 = 1, \dot{u}_2 = 0, \dot{v} = 0 \tag{9.5}$$

on a neighborhood of the origin of $\mathbb{R}^3$ (with coordinates u_1, u_2, v in $\mathbb{R}^3$). It is easy to see that we can choose H so that $H(z) = 0$, and the plane $\{v = 0\}$ is tangent to $H(W^u(p))$ at 0.

Consider the closed parallelepiped

$$D = \{(u_1, u_2, v) : 0 \le u_1 \le a, |u_2| \le b, |v| \le b\}$$

Choose $a, b > 0$ small enough so that $D \subset H(V)$. We now let $D_0 = D \cap \{u_1 = 0\}, D_1 = D \cap \{u_1 = a\}$. Evidently D_0, D_1 are transversal to trajectories of (9.5), hence, $H^{-1}(D_0), H^{-1}(D_1)$ are transversal to trajectories of (9.1). Choose $b > 0$ so small that the following statements are true:

(1) the component R_0 of $W^u(p) \cap H^{-1}(D_0)$ containing z is a smooth curve;

(2) the component R_1 of $W^s(q) \cap H^{-1}(D_1)$ containing $\varphi(a, z)$ is a smooth curve;

(3) for $x \in R_0,\ t < 0$, and for $x \in R_1,\ t > 0$

$$\varphi(t, x) \notin H^{-1}(D).$$

Consider smooth curves $Q_i = H(R_i),\ i = 0, 1$. As $H(W^u(p))$ is tangent to $\{v = 0\}$ at the origin, the curve Q_0 is given by $v = g_0(u_2)$, where $g_0 \in C^2$, $g_0(0) = 0$, $g_0'(0) = 0$. We supposed that $W^s(q)$ is nontransversal to $W^u(q)$ at z, hence $H(W^s(q))$ is also tangent to the plane $\{v = 0\}$ at the origin. It is easy to see that the structure of trajectories of (9.5) implies the following: the curve Q_1 is given by $v = g_1(u_2)$, where $g_1 \in C^2$, $g_1(0) = 0$, $g_1'(0) = 0$.

The main technical part of the proof is contained in the following statement (an analogue of this statement is used to prove Lemma 7.10).

Lemma 9.2. *Given $\delta, \varepsilon > 0$ there exists $\Delta > 0$ and a function $Y_\Delta(u_1, u_2, v)$ such that:*

(1) Y_Δ *equals to zero outside of the set*

$$D_\delta = \{(u_1, u_2, v) : 0 < u_1 < a, |u_2| < \delta, |v| < \delta\};$$

(2) *for any point* ξ *of the curve* $\{u_1 = 0, v = g_0(u_2), |u_2| < \Delta\}$ *there exists a segment of trajectory of the system*

$$\dot{u}_1 = 1, \dot{u}_2 = 0, \dot{v} = Y_\Delta(u_1, u_2, v) \tag{9.6}$$

beginning at ξ, *belonging to* $\overline{D}_\delta$ *and ending at a point of the curve* $\{u_1 = a, v = g_1(u_2), |u_2| < \Delta\}$;

(3) Y_Δ *is of class* C^2 *in* $H(V)$;

(4) *the* C^1*-distance between systems* (9.5) *and* (9.6) *is less than* ε.

Proof. Define the function $g(u_2) = g_1(u_2) - g_0(u_2)$ for $|u_2| < b$. Evidently

$$g(0) = 0, g'(0) = 0. \tag{9.7}$$

Fix $\Delta, \varepsilon, \delta > 0$ such that $0 < \Delta < \delta < b$. Consider functions $\alpha, \beta, \gamma \in C^\infty(\mathbb{R})$ such that

$\alpha(t) = 0$ for $t \in (-\infty, 0] \cup [a, +\infty)$,
$\alpha(t) = 1$ for $t \in [\frac{a}{3}, \frac{2a}{3}]$,
$0 < \alpha(t) < 1$ for $t \in (0, \frac{a}{3}) \cup (\frac{2a}{3}, a)$;
$\beta(t) = 0$ for $t \in (-\infty, -2\Delta] \cup [2\Delta, +\infty)$,
$\beta(t) = 1$ for $t \in [-\Delta, \Delta]$,
$0 < \beta(t) < 1$ for $t \in (-2\Delta, -\Delta) \cup (\Delta, 2\Delta)$;
$\gamma(t) = 0$ for $t \in (-\infty, -\delta] \cup [\delta, +\infty)$,
$\gamma(t) = 1$ for $t \in [-\frac{\delta}{2}, \frac{\delta}{2}]$,
$0 < \gamma(t) < 1$ for $t \in (-\delta, -\frac{\delta}{2}) \cup (\frac{\delta}{2}, \delta)$.

We can choose these functions so that

$$|\alpha'| \le \frac{4}{a}, |\beta'| \le \frac{2}{\Delta}, |\gamma'| \le \frac{3}{\delta}. \tag{9.8}$$

Let

$$A = \int_0^a \alpha(\tau) d\tau,$$

then

$$A > \frac{a}{3}.$$

Define Y_Δ by

$$Y_\Delta(u_1, u_2, v) = \frac{1}{A} g(u_2)\alpha(u_1)\beta(u_2)\gamma(v).$$

It follows immediately that $Y_\Delta \in C^2$ and that $Y_\Delta = 0$ outside of D_δ. Take Δ so small that for $|u_2| < 2\Delta$

$$|g_0(u_2)| < \frac{\delta}{2}, |g(u_2)| < \min\left(\frac{\varepsilon a}{6}, \frac{\varepsilon a^2}{72}, \frac{\varepsilon a\delta}{54}, \frac{\delta}{2}\right), \tag{9.9}$$

$$|g'(u_2)| < \frac{\varepsilon a}{144}. \tag{9.10}$$

The possibility of choice of Δ follows from (9.7). Consider for $|u_2^0| < \Delta$ the following functions

$$\begin{aligned} u_1(t) &= t, \\ u_2(t) &= u_2^0, \\ v(t) &= g_0(u_2^0) + \frac{g(u_2^0)}{A}\int_0^t \alpha(\tau)d\tau. \end{aligned} \tag{9.11}$$

We obtain from (9.9) that $|v(t)| < \delta$. It follows from the choice of Δ that $|u_2(t)| < \Delta < \delta$, so for $t \in [0, a]$, the graph of (9.11) belongs to $[0, a] \times D_\delta$. Substituting (9.11) in (9.6) we see that (9.6) is a solution of (9.6) for $t \in [0, a]$. As $u_1 = 0, u_2 = u_2^0, v = g_0(u_2^0)$ for $t = 0$, and $u_1 = a, u_2 = u_2^0, v = g_1(u_2^0)$ for $t = a$, the trajectory of this solution has a segment beginning at a point of Q_0, ending at a point of Q_1 and belonging to $\overline{D}_\delta$.

Let us estimate the function Y_Δ and its first derivatives. It follows from the definition of Y_Δ that it is enough to consider $|u_2| < 2\Delta$. We obtain from (9.8)–(9.10)

$$|Y_\Delta| < \frac{|g(u_2)|}{A} \le \frac{3}{a} \cdot \frac{\varepsilon a}{6} = \frac{\varepsilon}{2}, \tag{9.12}$$

$$\left|\frac{\partial Y_\Delta}{\partial u_1}\right| \le \frac{|g(u_2)|}{A}|\alpha'(u_1)| \le \frac{3}{a} \cdot \frac{\varepsilon a}{72} \cdot \frac{4}{a} = \frac{\varepsilon}{6}, \tag{9.13}$$

$$\begin{aligned} \left|\frac{\partial Y_\Delta}{\partial u_2}\right| &\le \frac{1}{A}(|g'(u_2)| + |g(u_2)| \cdot |\beta'(u_2)|) \\ &\le \frac{3}{a} \cdot \frac{\varepsilon a}{144} + \frac{3}{a} \cdot \frac{|g(u_2)|}{|u_2|} \cdot \frac{2|u_2|}{\Delta} \end{aligned}$$

$$\leq \frac{\varepsilon}{48} + \frac{12\Delta a\varepsilon}{144\Delta} < \frac{\varepsilon}{6}. \tag{9.14}$$

When we estimate the second term we take into account that the inequality (9.10) implies

$$\frac{|g(u_2)|}{|u_2|} \leq \frac{\varepsilon a}{144} \text{ for } 0 < |u_2| < 2\Delta.$$

Similarly we have

$$|\frac{\partial Y_\Delta}{\partial v}| \leq \frac{|g(u_2)|}{A}|\gamma'(v)| \leq \frac{3}{a} \cdot \frac{\varepsilon a\delta}{54} \cdot \frac{3}{\delta} = \frac{\varepsilon}{6}. \tag{9.15}$$

It follows from (9.12)–(9.15) that the C^1-distance between (9.5) and (9.6) is less than ε. □

Denote by Ξ the vector field of system (9.6). Consider system

$$\dot{x} = F^*(x) \tag{9.16}$$

defined in the following way: the diffeomorphism H^{-1} conjugates (9.6) in D and (9.16) in $H^{-1}(D)$. Let $w = (u_1, u_2, v)$. It was shown in the proof of Theorem 8.3 that

$$F^*(x) = \frac{\partial H^{-1}}{\partial w}(H(x))\,\Xi(H(x)) \tag{9.17}$$

(see (8.7)). Differentiating (9.17) with respect to x we obtain

$$\begin{aligned} \frac{\partial F^*}{\partial x}(x) = {} & \frac{\partial^2 H^{-1}}{\partial w^2}(H(x))\frac{\partial H(x)}{\partial x}\,\Xi(H(x)) \\ & + \frac{\partial H^{-1}}{\partial w}(H(x))\frac{\partial \Xi}{\partial w}(H(x))\frac{\partial H(x)}{\partial x} \end{aligned} \tag{9.18}$$

There exists $N > 0$ such that for $x \in H^{-1}(D)$ we have

$$\max\left(\left\|\frac{\partial H}{\partial x}(x)\right\|, \left\|\frac{\partial H^{-1}}{\partial w}(H(x))\right\|, \left\|\frac{\partial^2 H^{-1}}{\partial w^2}(H(x))\right\|\right) < N. \tag{9.19}$$

Hence, given a neighborhood U of system (9.1) there exists $\varepsilon > 0$ having the following property: if the C^1-distance between (9.5) and (9.6) is less than ε, then the corresponding system (9.16) is in U. We leave it to the reader to understand the meaning of all terms in (9.18) and to formalize the preceding arguments. Note that we use the assumption $F \in C^2$ to obtain (9.19).

Take $\varepsilon > 0$ mentioned above. Take $\delta < b$ and find a corresponding $\Delta > 0$, applying Lemma 9.2. It follows from the construction of the system (9.16) that p, q are trajectories of (9.16) and both $W^u(p)$ and $W^s(q)$ contain a "lune" S_Δ covered by trajectories of (9.16) intersecting the curve

$$H^{-1}(\{u_1 = 0, v = g_0(u_2), |u_2| < \Delta\}).$$

System (9.16) is topologically equivalent to the Kupka–Smale system (9.2). Denote h_0 as the corresponding topological equivalence. Then $h_0(W^u(p)) = W^u(p^*), h_0(W^s(q)) = W^s(q^*)$. Here p^*, q^* belong to the set of rest points and of closed trajectories of system (9.2). Hence, the intersection $W^u(p^*) \cap W^s(q^*)$ contains the "lune" $h_0(S_\Delta)$. On the other hand, it is easy to see that if we take a point $r \in W^u(P^*) \cap W^s(q^*)$ then, as $W^u(p^*), W^s(q^*)$ are transversal at r, there exists a neighborhood U_0 of r and a diffeomorphism $\nu : U_0 \to \mathbb{R}^3$ having the following property. The diffeomorphism ν maps neighborhoods of r in $W^u(p^*)$ and in $W^s(q^*)$ (in their inner topologies) onto 2-dimensional balls in two orthogonal planes of $\mathbb{R}^3$. Hence, $W^u(p^*) \cap W^s(q^*)$ contains no 2-dimensional "lunes." The contradiction we obtained proves Theorem 9.1. □

2. A similar result is true in the case of a structurally stable diffeomorphism.

Chapter 10

Homoclinic Point

1. Consider the autonomous system of differential equations (1.1) in $\mathbb{R}^n$ or on a closed smooth n-dimensional manifold M.

Let p be a hyperbolic rest point or a hyperbolic closed trajectory of system (1.1), and let $W^s(p), W^u(p)$ be the correpsonding stable and unstable manifolds. A point x_0 not belonging to the trajectory p and such that

$$x_0 \in W^u(p) \cap W^s(p) \tag{10.1}$$

is called a homoclinic point of the trajectory p. The point x_0 is sometimes called double-asymptotic to p, this name is due to

$$\varphi(t, x_0) \underset{t \to \pm\infty}{\longrightarrow} p.$$

The trajectory of a homoclinic point is called a homoclinic trajectory. If $W^u(p)$ and $W^s(p)$ are transversal at x_0 we say that x_0 is a transversal homoclinic point. Homoclinic points were mentioned in Section 3 of Chapter 6. It was shown there that if p is a hyperbolic rest point then there exist no transversal homoclinic trajectories of p.

It was Poincaré who first noticed that the existence of homoclinic points makes the structure of the set of trajectories very complicated. S. Smale, Yu. I. Neimark and others [16, 34, 38] studied in detail the structure of a neighborhood of transversal homoclinic trajectory.

In Section 2 of this chapter we show that a homoclinic point is nonwandering. In Section 3 we study the most visual case—a transversal homoclinic point for a diffeomorphism of the plane. Section 4 is devoted to the case of a 3-dimensional autonomous system of differential equations.

2. **Lemma 10.1.** *If x_0 is a homoclinic point of a hyperbolic rest point or of a hyperbolic closed trajectory of system* (1.1) *then x_0 is nonwandering.*

Proof. The point x_0 does not belong to the trajectory p and (10.1) is satisfied. Consider an arbitrary neighborhood U of x_0 and arbitrary $T > 0$. Choose an n-dimensional ball N in U such that $x_0 \in N$. As dim N equals to the dimension of the phase space, x_0 is a point of transversal intersection of N and $W^s(p)$. Consider the set

$$N_T = \bigcup_{t \geq T,\ x \in N} \varphi(t, x).$$

By Theorem 6.2, $W^u(p) \subset \overline{N_T}$; hence there exist points of the set N_T belonging to N (and consequently belonging to U). This proves that x_0 is nonwandering. □

3. Let $f : \mathbb{R}^2 \to \mathbb{R}^2$ be a diffeomorphism of class C^1. We assume that $p = 0$ is a saddle hyperbolic fixed point of f. We assume also that f is linear in a neighborhood of the origin. The last assumption allows us to avoid complicated estimates yet it slightly changes the principle ideas of the proof. So, in a neighborhood U of the origin with coordinates $x = (\xi, \eta)$ we have

$$f(\xi, \eta) = (\lambda\xi, \mu\eta) \tag{10.2}$$

where $0 < \lambda < 1, \mu > 1$. Evidently in this case

$$W^s_{loc}(0) = U \cap \{\eta = 0\},\ W^u_{loc}(0) = U \cap \{\xi = 0\}.$$

Suppose that q is a transversal homoclinic point of the saddle $p = 0$, i.e. q is a point of transversal intersection of $W^u(p), W^s(p), q \neq 0$. The manifolds $W^u(p), W^s(p)$ are invariant with respect to f, so by Lemma 6.1 the points $f^k(q)$, $k \in \mathbf{Z}$, are transversal homoclinic points of the saddle p. Since $f^k(q) \underset{t \to \pm\infty}{\longrightarrow} 0$ we can choose two points $q_u, q_s \in \{f^k(q)\}$ such that

$$q_u \in W^u_{loc}(0),\quad q_s \in W^s_{loc}(0).$$

There exists a positive integer m such that $q_s = f^m(q_u)$. Evidently $f^k(q_s) \in U$ for $k \geq 0$, $f^k(q_u) \in U$ for $k \leq 0$.

Consider arbitrary neighborhoods $U_1, \dots, U_{m-1}$ of the points $f(q_u), \dots, f^{m-1}(q_u)$. The set

$$\widetilde{V} = U \cup U_1 \cup \dots \cup U_{m-1}$$

is a neighborhood of the union of the trajectories of the saddle p and the point q. We call $\widetilde{V}$ an extended neighborhood of the homoclinic trajectory

$$\{f^k(q) : k \in \mathbf{Z}\}. \tag{10.3}$$

Theorem 10.1. *Given any neighborhood V of the point q_s, any extended neighborhood $\widetilde{V}$ of trajectory* (10.3) *and any positive m_0 there exists a periodic point r of the diffeomorphism f such that*
(1) $r \in V; f^k(r) \in \widetilde{V}, k \in \mathbf{Z}$;
(2) *the period of r is greater than m_0.*

Proof. For definiteness we assume that the neighborhood U is

$$U = \{(\xi, \eta) : |\xi| < R, |\eta| < R\}, \quad R > 0.$$

Let $q_s = (\xi_0, 0), q_u = (0, \eta_0)$, assume that $\xi_0 > 0, \eta_0 > 0$. Fix $a, b > 0$ so that the closed neighborhoods

$$\Pi_0 = \{(\xi, \eta) : |\xi - \xi_0| \le a, |\eta| \le a\},$$
$$\Pi_1 = \{(\xi, \eta) : |\xi| \le b, |\eta - \eta_0| \le b\}$$

of the points q_s, q_u are in U. Take an arbitrary neighborhood V of q_s and neighborhoods $U_1, \dots, U_{m-1}$ of the points $f(q_u), \dots, f^{m-1}(q_u)$. Suppose that a, b are so small that the inclusions

$$\Pi_0 \subset V, f^k(\Pi_1) \subset U_k, k = 1, \dots, m-1,$$

hold. Some additional restrictions on a, b are given below. The map f^m is a diffeomorphism of a neighborhood of q_u on a neighborhood of q_s. Suppose that $f^m(\xi, \eta) = (\xi_1, \eta_1)$ so that

$$\xi_1 = \xi_0 + g_1(\xi, \eta - \eta_0),$$
$$\eta_1 = g_2(\xi, \eta - \eta_0).$$

It follows from our assumptions that $g_1, g_2 \in C^1$ and that $g_1(0,0) = g_2(0,0) = 0$. Denote the Jacobi matrix

$$Df^m(q_u) = \begin{pmatrix} \frac{\partial g_1}{\partial \xi}(0,0) & \frac{\partial g_1}{\partial \eta}(0,0) \\ \frac{\partial g_2}{\partial \xi}(0,0) & \frac{\partial g_2}{\partial \eta}(0,0) \end{pmatrix} = \begin{pmatrix} \alpha_1 & \alpha_2 \\ \alpha_3 & \alpha_4 \end{pmatrix}.$$

The tangent space $T_{q_u}W^u(0)$ is the line $\{\eta = 0\}$, so the transversality of $W^u(0)$ and $W^s(0)$ at q_s is equivalent to the linear independence of the vectors

$$\begin{pmatrix} 1 \\ 0 \end{pmatrix}, \begin{pmatrix} \alpha_2 \\ \alpha_4 \end{pmatrix}.$$

The last condition is evidently equivalent to the inequality $\alpha_4 \neq 0$. Assume for definteness that

$$\alpha_4 > 0 \tag{10.4}$$

From the representation (10.2) of the diffeomorphism f it follows that if $\tilde{f}$ is the restriction of f on U then for $k > 0$ we have

$$\tilde{f}^k(\Pi_0) = \{(\xi, \eta) : |\xi - \lambda^k \xi_0| \leq a\lambda^k, |\eta| \leq a\mu^k\}.$$

Evidently, for a given $b > 0$, there exists $\kappa(b)$ such that for $k \geq \kappa(b)$ the set $\tilde{f}^k(\Pi_0) \cap \Pi_1$ is a nonempty rectangle with sides $\tilde{\sigma}_1, \tilde{\sigma}_2, \tilde{\sigma}_3, \tilde{\sigma}_4$ and its preimage

$$\tilde{\Pi}_k = \tilde{f}^{-k}\left(\Pi_1 \cap \tilde{f}^k(\Pi_0)\right)$$

is a nonempty rectangle with sides $\sigma_1, \sigma_2, \sigma_3, \sigma_4$ (see Figure 15). In a neighborhood of q_u we can represent f^m as

$$\xi_1 = \xi_0 + \alpha_1 \xi + \alpha_2(\eta - \eta_0) + \Xi(\xi, \eta - \eta_0),$$
$$\eta_1 = \alpha_3 \xi + \alpha_4(\eta - \eta_0) + H(\xi, \eta - \eta_0)$$

where

$$\frac{|\Xi(\xi, \eta - \eta_0)|}{|\xi| + |\eta - \eta_0|} \to 0, \frac{|H(\xi, \eta - \eta_0)|}{|\xi| + |\eta - \eta_0|} \to 0 \tag{10.5}$$

as $|\xi| + |\eta - \eta_0| \to 0$.

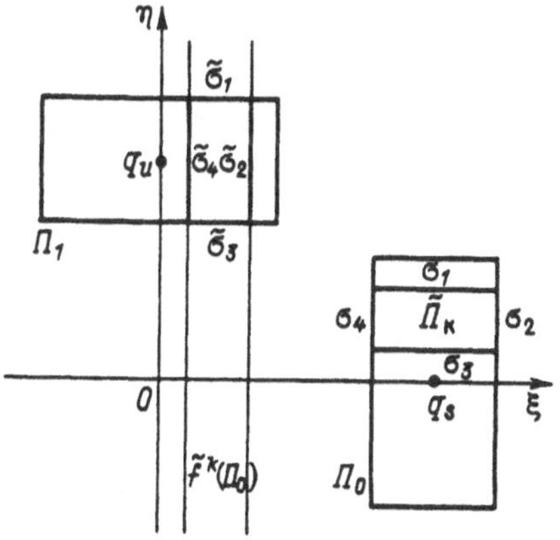

Figure 15

Take numbers $b > 0$, $k \geq \kappa(b)$ and a point $(\xi, \eta) \in \tilde{\sigma}_1$. The coordinates of the point are

$$\xi \in [\lambda^k(\xi_0 - a), \lambda^k(\xi_0 + a)], \eta = \eta_0 + b.$$

Let

$$\gamma_1(k, b, \theta) = \alpha_3 \lambda^k(\xi_0 + \theta a) + H(\lambda^k(\xi_0 + \theta a), b),$$

then for the point $(\xi_1, \eta_1) = f^m(\xi, \eta)$ there exists $\theta \in [-1, 1]$ such that

$$\eta_1 = \alpha_4 b + \gamma_1(k, b, \theta).$$

Similarly if $(\xi, \eta) \in \tilde{\sigma}_3$, there exists $\tilde{\theta} \in [-1, 1]$ such that

$$\eta_1 = -\alpha_4 b + \gamma_1(k, b, \tilde{\theta}).$$

The coordinates of a point $(\xi, \eta) \in \tilde{\sigma}_2$ are

$$\xi = \lambda^k(\xi_0 + a), \eta \in [\eta_0 - b, \eta_0 + b],$$

let

$$\gamma_2^{\pm}(k, b, \theta) = \alpha_1 \lambda^k(\xi_0 \pm a) + \alpha_2 \theta b + \Xi(\lambda^k(\xi_0 \pm a), \theta b).$$

Then for $(\xi, \eta) \in \tilde{\sigma}_2$ there exists $\theta \in [-1, 1]$ such that

$$\xi_1 = \xi_0 + \gamma_2^+(k, b, \theta),$$

for $(\xi, \eta) \in \tilde{\sigma}_4$ there exists $\tilde{\theta} \in [-1, 1]$ such that

$$\xi_1 = \xi_0 + \gamma_2^-(k, b, \tilde{\theta}).$$

It follows from the definition of the functions $\gamma_1, \gamma_2^{\pm}$ and from (10.5) that there exists $b_1 > 0$ having the following property: for any $b \leq b_1$, there exists $k_1 \geq \kappa(b)$ such that for $k \geq k_1$ and for $\theta \in [-1, 1]$ the inequalities

$$|\gamma_1(k, b, \theta)| < \frac{\alpha_4 b}{2}, \tag{10.6}$$

$$|\gamma_2^{\pm}(k, b, \theta)| < \frac{a}{2} \tag{10.7}$$

are true.

To complete the proof of Theorem 10.1 we need the following lemma from the theory of vector fields on the plane (its proof follows from Theorems 2.4, 3.1, and 4.1 in [10]).

Lemma 10.2. *Consider a continuous vector field*

$$w(\xi,\eta) = (w_1(\xi,\eta), w_2(\xi,\eta))$$

on a rectangle

$$\Pi = \{(\xi,\eta) : |\xi - \xi^0| \le R_0, |\eta - \eta^0| \le R_1\}.$$

Suppose that for $(\xi,\eta) \in \Pi$ *with* $|\xi - \xi^0| = R_0$

$$w_1(\xi,\eta)(\xi - \xi^0) > 0$$

and that for $(\xi,\eta) \in \Pi$ *with* $|\eta - \eta^0| = R_1$

$$w_2(\xi,\eta)(\eta - \eta^0) > 0$$

Then there exists a point $(\xi^*\eta^*) \in \Pi$ *such that* $w(\xi^*,\eta^*) = 0$

Now consider an arbitrary number m_0. Let $b = b_1$. Consider a rectangle

$$\tilde{\Pi}_k = \tilde{f}^{-k}\left(\Pi_1 \cap \tilde{f}^k(\Pi_0)\right)$$

for $k \ge \kappa(b_1)$ (see Figure 15). It is easy to see that the upper and lower sides of $\tilde{\Pi}_k$ are respectively

$$\sigma_1 = \{(\xi,\eta) : |\xi - \xi_0| \le a, \eta = \mu^{-k}(\eta_0 + b)\},$$
$$\sigma_3 = \{(\xi,\eta) : |\xi - \xi_0| \le a, \eta = \mu^{-k}(\eta_0 - b)\}.$$

We can choose a number $k_2 >$max (k_1, m_0) such that

$$\mu^{-k}(\eta_0 + b) < \frac{\alpha_4 b}{2} \tag{10.8}$$

for $k \ge k_2$. Consider a vector field $\tilde{w}(\xi,\eta) = f^{m+k}(\xi,\eta) - (\xi,\eta)$ on $\tilde{\Pi}_k$ for $k \ge k_2$. Let $\tilde{w} = (\tilde{w}_1, \tilde{w}_2)$. The field $\tilde{w}$ is evidently defined and continuous on $\tilde{\Pi}_k$. Define a field $w = (\tilde{w}_1, -\tilde{w}_2)$ on $\tilde{\Pi}_k$. This field is also continuous.

Take, for example, a point $(\xi,\eta) \in \sigma_1$. Then $f^k(\xi,\eta) \in \tilde{\sigma}_1$. If $f^{m+k}(\xi,\eta) = (\xi_1,\eta_1)$ then there exists $\theta \in [-1,1]$ such that

$$\eta_1 = \alpha_4 b + \gamma_1(k,b,\theta).$$

It follows from the equality $\eta = \mu^{-k}(\eta_0 + b)$ and from the inequalities (10.4), (10.6) and (10.8) that $\eta_1 > \eta$, so $\tilde{w}_1(\xi, \eta) > 0$. Similarly, $\tilde{w}_1 < 0$ on σ_3, $\tilde{w}_2 < 0$ on σ_2, $\tilde{w}_2 > 0$ on σ_4. Thus the field w satisfies the conditions of Lemma 10.2, so there exists a point $r = (\xi^*, \eta^*) \in \tilde{\Pi}_k$ such that $w(r) = 0$. It follows from the definition of the field w that $\tilde{w}(r) = 0$, so $f^{m+k}(r) = r$. The representation (10.2) of the diffeomorphism f in U shows that the points $r, f(r), \dots, f^k(r)$ belonging to U are distinct. Therefore, the period of the point r is greater than m_0.

It follows from our construction that point $r \in \Pi_0 \subset V$, and for any $k \in \mathbf{Z}$ either $f^k(r) \in U$ or $f^k(r) \in f^j(\Pi_1)$, $j = 1, \dots, m-1$, so the trajectory of point r belongs to the chosen extended neighborhood $\tilde{V}$. This completes the proof. □

4. Let γ be a saddle hyperbolic closed trajectory of period ω of an autonomous system (1.1) in $\mathbb{R}^3$. Then dim $W^u(\gamma) =$ dim $W^s(\gamma) = 2$. Let x_0 be a point of transversal intersection of $W^u(\gamma)$ and of $W^s(\gamma)$ not belonging to γ. Consider a 2-dimensional smooth disc S-transversal to γ and assume that $0 \in \gamma \cap S$. We denote by T the corresponding Poincaré map and by $W^s_{loc}(0), W^u_{loc}(0)$ the local stable manifold and the local unstable manifold of the saddle fixed point $0 \in S$ of T. Consider the points $q_0 \in W^s_{loc}(0) \cap \varphi(t, x_0), q_1 \in W^u_{loc}(0) \cap \varphi(t, x_0)$. By Theorem 1.1 there exists a diffeomorphism L of a neighborhood of the point q_1 in S on a neighborhood of the point q_0 in S (if we choose the points q_0, q_1 close enough to 0). Consider an arbitrary neighborhood U of the closed trajectory γ and choose q_0, q_1 belonging to U. Now consider an arbitrary neighborhood U_1 of the compact arc of the trajectory $\varphi(t, x_0)$ which joins the points q_1 and q_0. The union

$$\tilde{V} = U \cup U_1$$

is called an extended neighborhood of the homoclinic trajectory $\varphi(t, x_0)$.

Arguments similar to the proof of Theorem 10.2 show that for any neighborhood V of the point q_0 in S and for any m_0 we can find a point $r \in V$ and a number $k \geq m_0$ such that $L(T^k(r)) = r$. The transformations L, T are shifts along trajectories of system (1.1), so the point r is a point on a closed trajectory of the system. We can make the period of this closed trajectory as large as we want by choosing k large enough. The time in which the point q_0 reaches the point x_0 along the trajectory is finite. From the continuity of the flow of the system (1.1) it follows that the next theorem is true.

Theorem 10.2 *Given any neighborhood V_0 of a transversal homoclinic point x_0, any extended neighborhood $\widetilde{V}$ of the trajectory $\varphi(t, x_0)$ and any positive number ω_0 there exists a closed trajectory of system* (1.1) *which*

(1) *intersects V_0,*

(2) *belongs to $\widetilde{V}$,*

(3) *has minimal period greater than ω_0.*

5. It follows from Theorems 6.3 and 7.2 that the transversality of the intersection of $W^u(p)$ and of $W^s(p)$ for a hyperbolic closed trajectory p is preserved under C^1-small perturbations of the system. So the property of a system to have a transversal homoclinic point (and, consequently, to have a countable set of distinct closed trajectories) is open. That means the following: if a system has this property then all systems from its small neighborhood in X also have it.

It can be shown that there exist structurally stable systems having transversal homoclinic trajectories of saddle hyperbolic closed trajectories [38]. This proves the existence of structurally stable systems of differential equations with infinite sets of closed trajectories.

Chapter 11

Morse–Smale Systems

1. We mentioned in Chapter 3 the original definition of structural stability for autonomous systems of differential equations given by Andronov and Pontryagin in [1]. In this work Andronov and Pontryagin considered systems on a two-dimensional sphere, S^2 or on a disc $D^2 \subset \mathbb{R}^2$; it was supposed in the latter case that trajectories intersect the boundary of D^2 entering D^2 as t grows.

We consider here system (1.1) on S^2. Let p be a saddle hyperbolic rest point of (1.1). We call the stable and unstable manifolds $W^s(p)$ and $W^u(p)$ separatrices of p. In our case separatrices of p are one-dimensional.

We say that system (1.1) has a saddle-connection if there exists a point x that is different from rest points and two saddle hyperbolic rest points p_1, p_2 (not necessarily different) such that

$$\varphi(t,x) \underset{t\to-\infty}{\longrightarrow} p_1, \quad \varphi(t,x) \underset{t\to+\infty}{\longrightarrow} p_2.$$

Theorem 11.1 (The Andronov–Pontryagin Theorem [1]). *System (1.1) on S^2 is structurally stable if and only if:*

(1) *its rest points are hyperbolic;*

(2) *its closed trajectories are hyperbolic;*

(3) *there are no saddle-connections.*

Remark. We now formulate the Andronov–Pontryagin Theorem in terms of this book which differs from the terms of [1].

The necessity of the conditions of Theorem 11.1 for structural stability is a consequence of Theorem 9.1 (it is easy to see that according to the Corollary of Lemma 6.5, the transversality condition implies the absence of saddle-connections).

Any known proof of the sufficiency of the conditions of Theorem 11.1 is very long; we give no proofs here.

Suppose now that system (1.1) is structurally stable on S^2 and denote by P its set of rest points and of closed trajectories.

Theorem 11.2 (1) *The set P is finite.* (2) *The union of trajectories of the set P coincides with the nonwandering set of system* (1.1).

Proof. Consider a rest point p of system (1.1). If p is not a saddle then it is either attractive for the flow $\varphi(t,x)$ or attractive for the flow $\varphi(-t,x)$ (see Chapter 4). In each of these cases there exists a neigborhood U of p such that $\varphi(t,x) \to p$ for $x \in U$ as $t \to +\infty$ or as $t \to -\infty$. It follows from Lemma 4.6 that we can choose such a neighborhood U of p that in the first case $\varphi(t,x) \in U$ for $x \in U,\ t \geq 0$, and in the second case $\varphi(t,x) \in U$ for $x \in U,\ t \leq 0$.

If γ is a closed trajectory of (1.1), consider its Poincaré transformation T corresponding to a smooth curve Γ transversal to γ. As Γ is one-dimensional, the hyperbolic fixed point $q = \gamma \cap \Gamma$ of T is either attractive for T or attractive for T^{-1}. Hence, in the case of two-dimensional phase space for a hyperbolic closed trajectory γ we can find a neighborhood U of γ such that $\varphi(t,x) \to \gamma$ for $x \in U$ as $t \to +\infty$ or as $t \to -\infty$. Using arguments similar to the proof of Lemma 4.6 one can show that there exists a neighborhood U of γ such that $\varphi(t,x) \in U$ for $x \in U,\ t \geq 0$ in the first case, and $\varphi(t,x) \in U$ for $x \in U,\ t \leq 0$ in the second case.

It follows from Theorem 4.4 that for rest point p of (1.1) there exists a neighborhood containing no rest points different from p. The set of rest points is closed, so it is easy to see that the set of rest points of (1.1) is finite. Thus we can apply the Poincaré–Bendixson Theorem [6] to (1.1). It follows from this result that for any $x \in S^2$ we can describe the structure of the sets α_x and ω_x (see Section 3 of Chapter 2) in the following way. For the set ω_x (and also for the set α_x) one of the possibilities holds:

(1) ω_x is a rest point:

(2) ω_x is a closed trajectory;

(3) ω_x consists of rest points $p_1, \ldots, p_m$ and trajectories $\varphi(t,\xi)$ such that

$$\varphi(t,\xi) \underset{t\to-\infty}{\longrightarrow} p_i,\ \varphi(t,\xi) \underset{t\to+\infty}{\longrightarrow} p_j,\ i,j \in \{1,\ldots,m\}.$$

It follows from properties of rest points different from saddles that if such a rest point coincides with α_x for some x then it cannot coincide with ω_y for some y. Hence, if for some x the third possibility described in the Poincaré–Bendixson Theorem holds, then each rest point p_j, $1 \leq j \leq m$, is a saddle. In this case the set ω_x is a union of saddles and their separatrices, which contradicts the third statement of Theorem 11.1.

Consider the nonwandering set Ω of system (1.1). Every trajectory from the set P is a subset of Ω. Fix a point x. As it is shown, the sets α_x and ω_x are elements of P. It follows from the third statement of Theorem 11.1. that either α_x or ω_x is not a saddle rest point. Suppose that ω_x is not a saddle rest point. Consider two following possibilities:

(1) $x \in \omega_x$;

(2) $x \notin \omega_x$.

In the first case x belongs to the trajectory from P. In the second case, choose a neighborhood $U(\omega_x)$ such that $\varphi(t, \xi) \in U(\omega_x)$ for $\xi \in U(\omega_x)$, $t \geq 0$ (this is possible as ω_x is either an attractive rest point or an attractive closed trajectory). We can take $U(\omega_x)$ so small that

$$x \notin U(\omega_x).$$

There exists $T > 0$ such that $\varphi(T, x) \in U(\omega_x)$. It follows from the continuity of φ that there exists a neigborhood U_0 of x such that $U_0 \cap U(\omega_x) = \emptyset$ and $\varphi(T, \xi) \in U(\omega_x)$ for $\xi \in U_0$. We obtain from the properties of $U(\omega_x)$ that $\varphi(t+T, \xi) \in U(\omega_x)$ for $\xi \in U_0$, $t \geq 0$. Hence, if $x \notin \omega_x$, then $x \notin \Omega$. So we have proved the second statement of the theorem.

The set Ω is closed. Any trajectory p from P has a neighborhood U_p such that $U_p \cap U_q = \emptyset$ for different $p, q \in P$. As the union of trajectories of the set P is closed, we see that the set P is finite. □

2. Consider now a smooth closed n-dimensional manifold M and let $X = X^1(M)$. We say that a system (1.1) is a Morse–Smale system if:

(1) its nonwandering set Ω coincides with the union of hyperbolic rest points and a finite set of hyperbolic closed trajectories;

(2) the transversality condition holds.

In other words, system (1.1) is a Morse–Smale system if and only if (1.1) is a Kupka–Smale system having a finite set of closed trajectories.

Morse–Smale systems were introduced in [36] as a generalization of structurally stable autonomous two-dimensional systems. Indeed, it follows from Theorem 11.2 that if $M = S^2$, then system (1.1) is Morse–Smale if and only if statements (1)–(3) of Theorem 11.1 hold.

Defining the class of systems in [36] Smale supposed that, in addition to conditions (1) and (2) above, the following condition is satisfied: there do not exist any points x_0 and closed trajectories γ such that

$$x_0 \notin \gamma, \varphi(t, x_0) \underset{t \to \pm\infty}{\longrightarrow} \gamma.$$

It follows from Lemma 10.1 that this condition is a consequence of the condition (1).

J. Palis and S. Smale [20] proved the following theorem. It generalizes the sufficient conditions of structural stability given by Theorem 11.1. We do not give a proof of this result.

Theorem 11.3. *A Morse–Smale system is structurally stable.*

We prove the following result.

Theorem 11.4. *If the set of closed trajectories of a structurally stable system is finite, then this system is Morse–Smale.*

Proof. Suppose that the set of closed trajectories of system (1.1) is finite and that (1.1) is structurally stable. It follows from Theorem 9.1 that (1.1) is a Kupka–Smale system, hence its rests points and closed trajectories are hyperbolic. Applying Lemma 7.1 we see that the set of rest points is finite. Therefore the union of rest points and closed trajectories is compact. Define this union by Γ.

Let us show that Γ coincides with the nonwandering set Ω. To get a contradiction suppose that $\Omega \neq \Gamma$. As $\Gamma \subset \Omega$ we obtain that there exists $x_0 \in \Omega \backslash \Gamma$. Denote by U a neighborhood of system (1.1) described in the definition of structural stability. It follows from Theorem 8.2 that there exists in U a system

$$\dot{x} = \widetilde{F}(x) \tag{11.1}$$

such that the union of rest points and of closed trajectories of (11.1) is dense in $\Omega(\widetilde{F})$. Let h be a topological equivalence of systems (1.1) and (11.1). It follows from Lemma 2.2 that $h(x_0) \in \Omega(\widetilde{F})$. As Γ is compact, the point $h(x_0)$ has a neighborhood W such that $W \cap h(\Gamma) = \emptyset$. It is easy to see that $h(\Gamma)$ coincides with the union of rest points and of closed trajectories of system (11.1). The contradiction we obtained completes the proof. □

3. S. Smale studied in [36] the so called "geometric theory of Morse–Smale systems." The two following statements are basic to this theory:

(1) The manifold M is a union of unstable manifolds of rest points and closed trajectories;

(2) unstable manifolds of rest points and closed trajectories are submanifolds of M.

We showed in Chapter 4 that the unstable manifold of a hyperbolic point is not necessarily a submanifold of the phase space.

To prove the basic statements of this theory of Smale's we study the limit sets of stable and unstable manifolds.

Let p be a rest point or a closed trajectory of a Morse–Smale system (1.1).

Define the following set:

$$\omega(W^u(p)) = \left\{ \begin{array}{l} y = \lim\limits_{k\to\infty} \varphi(t_k, x_k) : x_k \in W^u(p),\ t_k \underset{k\to\infty}{\longrightarrow} +\infty, \\ \text{sequences } x_k \text{ have no limit points on } p \end{array} \right\}.$$

We call the set $\omega(W^u(p))$ the ω-limit set of the unstable manifold $W^u(p)$.

We define similarly (taking $x_k \in W^s(p), t_k \underset{k\to\infty}{\longrightarrow} -\infty$) the set $\alpha(W^s(p))$, called the α-limit set of $W^s(p)$. It is easy to see that the sets $\omega(W^u(p)), \alpha(W^s(p))$ are compact and invariant with respect to the flow φ.

Let p be a rest point of system (1.1). We defined in Section 3 of Chapter 4 the parametrizing sphere Σ for $W^s(p)$. Denote now this sphere by $\sum_p^s$, and denote an analogous parametrizing sphere for $W^u(p)$ by $\sum_p^u$. We can describe the sets $\omega(W^u(p)), \alpha(W^s(p))$ using these spheres. We prove the statement we need only for $\omega(W^u(p))$.

Lemma 11.1.

$$\omega(W^u(p)) = \{y = \lim_{k\to\infty} \varphi(t_k, \sigma_k) : \sigma_k \in \sum_p^u, t_k \underset{k\to\infty}{\longrightarrow} +\infty\}. \tag{11.2}$$

Proof. Denote by Q the set being the right side of equality (11.2). For any sequence $\sigma_k \in \sum_p^u$ the point p is not a limit point, hence

$Q \subset \omega(W^u(p))$. To prove $\omega(W^u(p)) \subset Q$ consider $y \in \omega(W^u(p))$, $y = \lim\limits_{k\to\infty} \varphi(t_k, x_k)$. It follows from the definition of $\omega(W^u(p))$ that only a finite set of points x_k can coincide with p, and we suppose that $x_k \neq p$ for any k. Any trajectory $\varphi(t, x_k)$ has a unique point of intersection with $\sum_p^u$. Let $x_k = \varphi(\tau_k, \sigma_k)$, $\sigma_k \in \sum_p^u$. We obtain from Lemma 4.6 that the sequence τ_k is bounded from below. Indeed, if there exists a subsequence $\tau_{k_l} \to -\infty$, then by Lemma 4.6 the distance between $\varphi(t, \sigma_k)$ and p is bounded by $d \exp(\mu t)$ for $t \in [\tau_{k_l}, 0)$, where $\mu > 0$, so that the distances between $x_{k_l} = \varphi(\tau_{k_l}, \sigma_{k_l})$ and p tend to zero as $k_l \to \infty$.

As τ_k is bounded from below, we have $t_k + \tau_k \underset{k\to\infty}{\longrightarrow} +\infty$, and $y = \lim \varphi(t_k + \tau_k, \sigma_k)$, so $y \in Q$. □

Remark. There exist analogous constructions for a hyperbolic closed trajectory p, in which case parametrizing manifolds are more complicated than spheres.

Lemma 11.2. *For any $x \in M$ there exist nonwandering trajectories p, q such that*

$$x \in W^u(p) \cap W^s(q).$$

Proof. Consider the sets α_x and ω_x-the α-limit set and the ω-limit set of the trajectory $\varphi(t, x)$. As M is compact, the sets α_x and ω_x are non-empty and connected. It follows from Lemma 2.1 that $\alpha_x, \omega_x \subset \Omega$. As system (1.1) is Morse–Smale, the set Ω coincides with a union of a finite set of rest points and a finite set of closed trajectories. It is evident now that there exist trajectories p, q in Ω such that $p = \alpha_x$, $q = \omega_x$. It follows from Lemma 4.7 (and from the analogous statement for closed trajectories) that

$$x \in W^u(p),\ x \in W^s(q). \square$$

We defined in Chapter 6 the relation $p \to q$ for pairs p, q of rest points and hyperbolic trajectories: we write $p \to q$ if there exists a point $x \in W^u(p) \cap W^s(q)$ such that $x \notin p \cup q$. It is easy to see that for a Morse–Smale system $p \to q$ means the following: there exists a point $x \notin \Omega$ such that

$$\varphi(t, x) \underset{t\to-\infty}{\longrightarrow} p,\ \varphi(t, x) \underset{t\to+\infty}{\longrightarrow} q.$$

Lemma 11.3. *If $p \to q$ then $W^u(q) \subset \omega(W^u(p))$.*

Proof. Consider a point $x \in W^u(p) \cap W^s(q)$ such that $x \notin p \cup q$. Consider a closed smooth disc N in $W^u(p)$ such that $N \cap p \neq \emptyset$ and N is transversal to $W^s(q)$ at x. It was shown in Theorem 6.2 that for any $T > 0$ we have

$$W^u(q) \subset \overline{\bigcup_{t \geq T, x \in N} \varphi(t, x)}. \tag{11.3}$$

It is easy to see now that (11.3) implies $W^u(q) \subset \omega(W^u(p))$. □

It follows from Lemma 6.5 and from Lemma 10.1 that for a non-wandering trajectory p of a Morse–Smale system the relationship $p \to p$ is impossible. Using Theorem 6.4 we obtain the following statement.

Lemma 11.4. *For any chain*

$$p_1 \to p_2 \to \dots \to p_m, \; p_i \in \Omega,$$

the trajectories $p_1, \dots, p_m$ are distinct.

Lemma 11.5. *If for $p, q \in \Omega$*

$$\omega(W^u(p)) \cap W^u(q) \neq \emptyset,$$

then $p \to q$.

Proof. Suppose for definiteness that q is a rest point. Take a point $x \in \omega(W^u(p)) \cap W^u(q)$. Find a neighborhood U_1 of the point q and take coordinates (ξ, η) in U_1 such that q is the origin in these coordinates and the inequalities (4.45)and (4.46) hold.

As $\omega(W^u(p)), W^u(q)$ are invariant and as

$$\varphi(t, x) \underset{t \to -\infty}{\longrightarrow} q,$$

we can find a point $x_0 \in U_1 \cap \omega(W^u(p)) \cap W^u(q)$. Consider a sequence $x_m \in W^u(p)$ such that $x_m \to x_0$ as $m \to \infty$. We suppose that $x_m \in U_1$. As

$$W^u_{loc}(q) = U_1 \cap \{\xi = 0\}$$

the point x_0 has coordinates $(0, \eta_0)$. Let $x_m = (\xi_m, \eta_m)$. Then $\xi_m \to 0$, $\eta_m \to \eta_0$ as $m \to \infty$. We denote by $(\xi(t, \tilde{\xi}, \tilde{\eta}), \eta(t, \tilde{\xi}, \tilde{\eta}))$, the trajectory in U_1 with initial conditions $(0, \tilde{\xi}, \tilde{\eta})$. Take $b > 0$ so small that the sphere

$$Q_b = \{(\xi, \eta) : |\xi| = b, \eta = 0\}$$

belongs to U_1. We have $|\xi_m| < b$ for large m. As

$$|\xi(t, \xi_m, \eta_m)| \geq e^{-\mu t}|\xi_m|$$

for $t < 0$ (while the trajectory is in U_1), there exists $\tau_m < 0$ such that

$$(\xi(t, \xi_m, \eta_m), \eta(t, \xi_m, \eta_m)) \in U_1, \; t \in [\tau_m, 0],$$

$$|\xi(t, \xi_m, \eta_m)| < b, \; t \in (\tau_m, 0],$$

$$|\xi(\tau_m, \xi_m, \eta_m)| = b.$$

It is easy to see that $\tau_m \to -\infty$ as $m \to \infty$. Taking into account that

$$|\eta(t, \xi_m, \eta_m)| \leq e^{\mu t}|\eta_m|, \; t \leq 0,$$

we obtain that $|\eta(\tau_m, \xi_m, \eta_m)| \to 0$ for $m \to \infty$. Let x^* be a limit point of the sequence $\zeta_m = (\xi(\tau_m, \xi_m, \eta_m), \eta(\tau_m, \xi_m, \eta_m))$. It follows from our considerations that $x^* \in Q_b \subset W^s_{loc}(q)$. The points ζ_m belong to the trajectories of points $x_m \in W^u(p)$, hence there exists $\sigma_m \in \sum_p^u$ and θ_m such that $\zeta_m = \varphi(\theta_m, \sigma_m)$. The following are two possibilities.

If the sequence θ_m has a subsequence bounded from above then it has a convergent subsequence $\theta_{m_l} \to t^*$. The equality $t^* = -\infty$ is impossible, as in this case $x^* = p$. The parametrizing manifold $\sum_p^u$ is compact. Let σ^* be a limit point of σ_{m_l}. It follows from the continuity of the flow φ that

$$x^* = \varphi(t^*, \sigma^*) \in W^u(p),$$

and hence $p \to q$.

If $\theta_m \to +\infty$ as $m \to \infty$, then $x^* \in \omega(W^u(p))$. It follows from Lemma 11.2 that there exists a trajectory q_1 in Ω such that $x^* \in W^u(q_1)$, then

$$q_1 \to q, \; \omega(W^u(p)) \cap W^u(q_1) \neq \emptyset.$$

Repeating the arguments with the pair p, q_1 we obtain that either $p \to q_1$ (then $p \to q$ by Theorem 6.4) or there exists a trajectory q_2 in Ω such that

$$q_2 \to q_1, \; \omega(W^u(p)) \cap W^u(q_2) \neq \emptyset$$

and so on. It follows from Lemma 11.4 that the length of the chain $q \leftarrow q_1 \leftarrow q_2 \leftarrow \dots$ arising from our construction is not more than the number of trajectories in Ω. Therefore there exists a trajectory q_m in Ω such that $p \to q_m \to \dots \to q_1 \to q$, and then $p \to q$. □

The following statement is a consequence of Lemma 11.3 and of Lemma 11.5.

Theorem 11.5. *Let p, q be trajectories in Ω. Three following statements are equivalent:*
(1) $p \to q$;
(2) $\omega(W^u(p)) \supset \overline{W^u(q)}$;
(3) $\omega(W^u(p)) \cap W^u(q) \neq \emptyset$.

Remark. It is well-known that the ω-limit set of a trajectory consists of complete trajectories. Theorem 11.5 shows that for Morse–Smale systems the following generalization of the result is true: the ω-limit set of the unstable manifold of a trajectory in Ω consists of complete unstable manifolds of trajectories in Ω.

Lemma 11.6. *Let p be a trajectory in Ω. Then*

$$W^u(p) \cap \omega(W^u(p) = \emptyset.$$

Proof. If $W^u(p) \cap \omega(W^u(p)) \neq \emptyset$ then by Theorem 11.5 we have $p \to p$, and that is impossible.

Theorem 11.6. *Let p be a trajectory in Ω. Then $W^u(p)$ is a submanifold of M.*

Proof. We consider the case of a rest point p. Consider a point $x_0 \in W^u(p)$. To prove the theorem it is enough to show that there exists a neighborhood U of x_0 in M and a diffeomorphism $f : U \to \mathbb{R}^n$ such that $f(U) = \mathbb{R}^n$ and $f(U \cap W^u(p)) = \mathbb{R}^k \subset \mathbb{R}^n$.

We begin with the case $x_0 \neq p$. Let Σ be a parametrizing sphere in $W^u(p)$. There exists a unique pair (t_0, σ_0), $t_0 \in \mathbb{R}$, $\sigma_0 \in \Sigma$, such that $x_0 = \varphi(t_0, \sigma_0)$. It was shown in Chapter 4 that $W^u_{loc}(p)$ is the image of a diffeomorphism of a smooth disc. Therefore there exists a neighborhood U_0 of σ_0 in M and a diffeomorphism $h : U_0 \to \mathbb{R}^n$ such that $h(U_0) = \mathbf{R}^n$ and if $V = U_0 \cap W^u_{loc}(p)$ then $h(V) = \mathbf{R}^k \subset \mathbf{R}^n$. Define the diffeomorphism $\tau : \tau(x) = \varphi(-t_0, x)$, then $\sigma_0 = \tau(x_0)$. Let us show that if the neighborhood U_0 is small enough then

$$\tau^{-1}(V) = W^u(p) \cap \tau^{-1}(U_0). \tag{11.4}$$

If (11.4) holds then the neighborhood $U = \tau^{-1}(U_0)$ of x_0 has the desired properties: $h(\tau(U)) = \mathbf{R}^n$ and $h(\tau(W^u(p) \cap U)) = \mathbf{R}^k \subset \mathbf{R}^n$.

To get a contradiction we suppose that any neighborhood of x_0 contains points of $W^u(p)$ not belonging to $\tau^{-1}(V)$. Take a sequence $x_m \underset{m\to\infty}{\longrightarrow} x_0$ of points having this property. Evidently $x_m \neq p$ for large m. Then there exist $t_m \in \mathbf{R}$ and $\sigma_m \in \Sigma$ such that $x_m = \varphi(t_m, \sigma_m)$. Note that the sequence t_m is bounded from below (if there exists a subsequence $t_{m_l} \to -\infty$ then $x_{m_l} \to p$).

If $t_m \to +\infty$ as $m \to \infty$ then $x_0 \in \omega(W^u(p))$ and we get a contradiction with Lemma 11.6.

If t_m has a subsequence t_{m_l} bounded from above, consider a limit point (t', σ') of (t_{m_l}, σ_{m_l}). It follows from the uniqueness of a limit and from the continuity of φ that

$$x_0 = \varphi(t_0, \sigma_0) = \varphi(t', \sigma').$$

As $\varphi(t, x_0)$ has a unique point of intersection with Σ, we have $t' = t_0$, $\sigma' = \sigma_0$. We conclude from these equalities that $x_{m_l} \in \tau^{-1}(V)$ for large m_l, which leaves a contradiction as to the choice of x_{m_l}.

If $x_0 = p$, there exist a neighborhood U_0 of p and a diffeomorphism h such that $h(U_0) = \mathbf{R}^n$, $h(U_0 \cap W^u_{loc}(p)) = \mathbf{R}^k \subset \mathbf{R}^n$. Let

$$V = U_0 \cap W^u_{loc}(p).$$

Let us show that if the neighborhood U_0 is small enough then

$$W^u(p) \cap U_0 = V.$$

To get a contradiction suppose that there exists a sequence $x_m \in W^u(p)$ such that $x_m \to p$ as $m \to \infty$ and $x_m \notin V$. Evidently $x_m \neq p$; take $t_m \in \mathbf{R}$ and $\sigma \in \Sigma$ such that $x_m = \varphi(t_m, \sigma_m)$. It is easy to see that the sequence t_m is bounded from below. If t_m contains a subsequence bounded from above then there exists t', σ' such that $p = \varphi(t', \sigma')$ contradicting the fact that $\varphi(t, p) = p$ for $t \in \mathbf{R}$. The case $t_m \to +\infty$ is studied as above. □

It follows from Lemma 11.2 that

$$M = \bigcup_{p \subset \Omega} W^s(p) = \bigcup_{p \subset \Omega} W^u(p). \tag{11.5}$$

It was shown in Theorem 11.6 that $W^u(p)$ is a submanifold of M for any trajectory $p \subset \Omega$. Similar arguments prove that $W^s(p)$ are also submanifolds of M. Each $W^u(p), W^s(p)$ is either an embedded image of $\mathbb{R}^k$ (see Chapter 4) or an embedded image of a fibering over S^1 with fibers $\mathbb{R}^k$ (see Chapter 5). So, the manifold M is represented as a union of submanifolds having relatively simple topology. The representation (11.5) allowed the obtaining of important results connecting the structure of the set of trajectories of a Morse–Smale system on M with topological properties of M (see, for example, [4]).

4. Let f be a diffeomorphism of a closed smooth manifold M. We say that f is a Morse–Smale diffeomorphism if

(1) the nonwandering set Ω of f coincides with a finite set of hyperbolic periodic points;

(2) the transversality condition holds.

Analogues of results of Sections 2 and 3 are valid for Morse–Smale diffeomorphisms.

Chapter 12

Hyperbolic Sets

1. We studied in Chapters 4 and 5 hyperbolic rest points and hyperbolic closed trajectories of autonomous systems of differential equations, and also hyperbolic periodic points of diffeomorphisms. Now we are going to study hyperbolic sets. The notions of a hyperbolic set generalize those mentioned above.

Let us begin with the case of a diffeomorphism. Consider a diffeomorphism f of class C^1 of a smooth closed n-dimensional manifold M. For $x \in M$ we denote by

$$Df(x) : T_xM \to T_{f(x)}M$$

the derivative of f at x. Let d be a Riemannian metric on M and let $|v|$ be the corresponding norm of $v \in T_xM$.

Consider a set $I \subset M$. We say that I is a hyperbolic set of f (or we simply say: I is hyperbolic), if:

(1) I is compact and invariant with respect to f;

(2) there exist $C > 0$ and $\lambda \in (0,1)$ having the following property: for any $p \in I$ there exist two linear subspaces $L^+(p)$ and $L^-(p)$ of T_pM such that:

$$\text{(a)} \quad L^+(p) + L^-(p) = T_pM; \tag{12.1}$$

$$\text{(b)} \quad Df(p)L^+(p) = L^+(f(p)),$$

$$Df(p)L^-(p) = L^-(f(p)); \tag{12.2}$$

$$\text{(c)} \quad |Df^k(p)v| \le C\lambda^k\,|v| \text{ for } v \in L^+(p),\ k \ge 0; \tag{12.3}$$

$$\text{(d)} \quad |Df^k(p)v| \le C\lambda^{-k}|v| \text{ for } v \in L^-(p),\ k \le 0. \tag{12.4}$$

Note that in this definition the constants C, λ are the same for any point p of I. We write in this case $I \in H(C, \lambda)$.

It is easy to see that if p is a hyperbolic periodic point of p of period m then the set $\{p, f(p), \dots, f^{m-1}(p)\}$ is hyperbolic. For that set $L^+(p)$ is spanned by the eigenvectors of $Df^m(p)$ corresponding to eigenvalues λ_j of $Df^m(p)$ such that $|\lambda_j| < 1$, and $L^-(p)$ is spanned by the eigenvectors of $Df^m(p)$ corresponding to eigenvalues λ_j of $Df^m(p)$ such that $|\lambda_j| > 1$.

Remark. In the definition given above we require that $Df(p)$ maps $L^+(p)$ onto $L^+(f(p))$, and $L^-(p)$ onto $L^-(f(p))$ (condition (12.2)). Let us show that instead of this condition we can require the following to be true: for any $p \in I$ the dimensions of $L^+(f^k(p))$ and of $L^-(f^k(p))$ are constant with respect to $k \in \mathbf{Z}$.

To get a contradiction suppose that for some $p \in M$ and $\kappa \in \mathbf{Z}$ we have

$$Df^\kappa(p)L^+(p) \neq L^+(q)$$

where $q = f^\kappa(p)$. Let $\dim L^+(f^k(p)) = l$ for $k \in \mathbf{Z}$. As Df^κ is a nonsingular linear map,

$$\widetilde{L}^+(p) = Df^{-\kappa}(q)L^+(q) \neq L^+(p).$$

For any $v \in \widetilde{L}^+(p)$ we have $Df^\kappa(p)v \in L^+(q)$, it follows from (12.3) that

$$|Df^k(p)v| \leq C\lambda^{k-\kappa}|Df^\kappa(p)v| \text{ for } k \geq \kappa.$$

Hence there exists $C_1 \geq C$ such that

$$|Df^k(p)v| \leq C_1\lambda^k|v| \tag{12.5}$$

for $v \in \widetilde{L}^+(p)$, $k \geq \kappa$.

The subspaces $\widetilde{L}^+(p)$ and $L^+(p)$ of T_pM have equal dimensions and do not coincide. Find a vector v_0, $|v_0| = 1$ that belongs to $\widetilde{L}^+(p) \backslash L^+(p)$. Let $\{v_0\}$ be the subspace of T_pM spanned by v_0 and let

$$\widehat{L}^+(p) = \{v_0\} + L^+(p).$$

The angle between v_0 and $L^+(p)$ is nonzero. Hence there exists $a > 0$ such that for any $w \in \widehat{L}^+(p)$, $|w| = 1$, the decomposition $w = a_1v_0 + a_2v_1$ where $v_1 \in L^+(p)$, $|v_1| = 1$ implies the inequalities $|a_1| \leq a$, $|a_2| \leq a$. Therefore it follows from (12.3) and (12.5) that there exists $C_2 \geq C_1$ having the following property: for $k \geq 0$ and for $w \in \widehat{L}^+(p)$

$$|Df^k(p)w| \leq C_2\lambda^k|w|. \tag{12.6}$$

Take $m > 0$ such that

$$C_2\lambda^m < \frac{1}{2}.$$

Let $r = f^m(p)$. As dim $L^-(f^k(p)) = \text{const}$, we obtain from (12.1) that

$$\dim Df^{-m}(r)L^-(r) = \dim L^-(p) \geq n - l.$$

As dim $\widehat{L}^+(p) = l + 1$, there exists a vector w_0 such that

$$w_0 \in \widehat{L}^+(p) \cap Df^{-m}(r)L^-(r),\ |w_0| = 1.$$

Then

$$1 = |w_0| = |Df^{-m}(r)Df^m(p)w_0| \leq \frac{|Df^m(p)w_0|}{2} \leq \frac{|w_0|}{4} = \frac{1}{4}.$$

We take into account (12.4) and $C\lambda^m \leq \frac{1}{2}$ in the first inequality, and (12.6) in the second inequality. The contradiction we obtained completes the proof of the remark.

The general notion of a hyperbolic set was introduced by D.V. Anosov [2]. Anosov studied dynamical systems such that the manifold M is a hyperbolic set. He called such systems (Y)-systems and now they are usually called Anosov systems.

We prove in Section 2 of this chapter that a diffeomorphism is structurally stable on a hyperbolic set (Theorem 12.8). It follows from Theorem 12.8 that Anosov diffeomorphisms are structurally stable (see Theorem 12.9).

Let us describe now some simple properties of hyperbolic sets. Consider a point p of a hyperbolic set $I \in H(C, \lambda)$. Take $k > 0$ and $q = f^k(p)$ and consider a vector $v \in L^-(p)$. It follows from (12.2) that $Df^k(p)v \in L^-(q)$. Using (12.4) we obtain

$$|v| = |Df^{-k}(q)Df^k(p)v| \leq C\lambda^k|Df^k(p)v|.$$

Hence for $v \in L^-(p)$ and for $k > 0$ we have

$$|Df^k(p)v| \geq \frac{\lambda^{-k}}{C}|v|.$$

We obtain similarly that

$$|Df^k(p)v| \geq \frac{\lambda^k}{C}|v|$$

for $v \in L^+(p)$ and for $k < 0$. It follows immediately from these inequalities that the sum in (12.1) is direct. Consider

$$v \in L^+(p) \cap L^-(p).$$

Take $k > 0$. Then

$$C\lambda^k |v| \geq |Df^k(p)v| \geq \frac{\lambda^{-k}}{C}|v|.$$

Hence,

$$\left(C\lambda^k - \frac{\lambda^{-k}}{C}\right)|v| \geq 0.$$

As

$$C\lambda^k - \frac{\lambda^{-k}}{C} \underset{k \to +\infty}{\longrightarrow} -\infty,$$

we obtain from the last inequality that $v = 0$.

As $T_pM = L^+(p) \oplus L^-(p)$ for any $v \in T_pM$ there exists a unique $v^+ \in L^+(p)$, $v^- \in L^-(p)$ such that $v = v^+ + v^-$. Define the linear map $R_p : T_pM \to T_pM$ by: $R_p v = v^+$. This map projects T_pM onto $L^+(p)$ parallel to $L^-(p)$.

Denote by $\sphericalangle(L^+(p), L^-(p))$ the least angle between vectors v_1, v_2 such that $v_1 \in L^+(p)$, $|v_1| = 1$, $v_2 \in L^-(p)$, $|v_2| = 1$.

As I is compact there exists $N > 0$ such that

$$\|Df(p)\| \leq N, \ \|Df^{-1}(p)\| \leq N \text{ for } p \in I. \tag{12.7}$$

Lemma 12.1. *There exists $\alpha_0 = \alpha_0(N, C, \lambda)$ such that for any $p \in I$*

$$\sphericalangle(L^+(p), L^-(p)) \geq \alpha_0. \tag{12.8}$$

Proof. Take vectors $v^+ \in L^+(p), v^- \in L^-(p), |v^+| = |v^-| = 1$, and define

$$\Delta(k) = Df^k(p)(v^- - v^+).$$

It follows from (12.7) that

$$|\Delta(k)| \leq N^k |\Delta(0)|, \ k \geq 0. \tag{12.9}$$

Let us estimate $|\Delta(k)|$ from below:

$$|\Delta(k)| \geq |Df^k(p)v^-| - |Df^k(p)v^+| \geq \frac{\lambda^{-k}}{C} - C\lambda^k.$$

There exists $k_0 = k_0(C, \lambda)$ such that $|\Delta(k_0)| \geq 1$. We obtain from (12.9) that

$$|v^- - v^+| = |\Delta(0) \geq N^{-k_0},$$

which completes the proof. □

Corollary. *There exists $\alpha_1 = \alpha_1(N, C, \lambda)$ such that for $p \in I$ we have*

$$\|R_p\| \leq \alpha_1,\ \|E_p - R_p\| \leq \alpha_1;$$

here E_p is the identity mapping $T_pM \to T_pM$.

To prove the corollary take

$$\alpha_1 = \frac{1}{\sin \alpha_0}.$$

Consider a subset M_0 of the manifold M. Consider a linear subspace $L(p)$ of T_pM for any $p \in M_0$. Using the structure of the manifold TM (see Section 4 in Chapter 1) we can define

$$\lim_{k\to\infty} L(p_k) = \{v \in T_pM : v = \lim_{k\to\infty} v_k,\ v_k \in T_{p_k}M\}$$

for a sequence $p_k \in M_0$ such that $p_k \underset{k\to\infty}{\longrightarrow} p$. We say that the family $L(p)$ is continuous at $p_0 \in M_0$ if for any sequence $p_k \in M_0, p_k \underset{k\to\infty}{\longrightarrow} p_0$,

$$\lim_{k\to\infty} L(p_k) = L(p_0).$$

Lemma 12.2. *Let I be a hyperbolic set. Then the families $L^+(p)$, $L^-(p)$ are continuous at any $p \in I$.*

Proof. Assume that $I \in H(C, \lambda)$. Consider a sequence $p_k \in I$ such that $p_k \underset{k\to\infty}{\longrightarrow} p \in I$. Let us prove the following: if for a sequence of vectors $v_k \in L^+(p_k)$, $|v_k| = 1$, we have $\lim_{k\to\infty} v_k = v$ then

$$v \in L^+(p). \tag{12.10}$$

To get a contradiction suppose that

$$v = v^+ + v^-, \ v^+ \in L^+(p), \ v^- \in L^-(p), \ v^- \neq 0.$$

Take $k_0 > 0$ such that

$$C\lambda^{k_0} < \frac{1}{3}, \ |Df^{k_0}(p)v^+| < \frac{1}{3}, \ |Df^{k_0}(p)v^-| > 1. \tag{12.11}$$

The first inequality in (12.11) implies

$$|Df^{k_0}(p_k)v_k| < \frac{1}{3}.$$

As $Df^{k_0}(r)w$ is continuous with respect to r and w we have

$$|Df^{k_0}(p)v| \le \frac{1}{3}.$$

Now using (12.11) we obtain that

$$|Df^{k_0}(p)v| \ge |Df^{k_0}(p)v^-| - |Df^{k_0}(p)v^+| > \frac{2}{3}.$$

The contradiction we obtained proves (12.10). It follows from (12.10) that

$$\widetilde{L}^+(p) = \lim_{k\to\infty} L^+(p_k) \subset L^+(p). \tag{12.12}$$

We can prove similarly that

$$\widetilde{L}^-(p) = \lim_{k\to\infty} L^-(p_k) \subset L^-(p).$$

Consider the sequence dim $L^+(p_k)$. There exists s such that dim $L^+(P_{k_m}) = s$ for an infinite subsequence p_{k_m} of p_k. Introduce coordinates in a neighborhood of (p, T_pM) in TM using the map β described in Section 4 of Chapter 1. Choose an orthonormal basis $v^1_{k_m}, \dots, v^s_{k_m}$ of $L^+(p_{k_m})$. We can suppose that there exists

$$v^i = \lim_{k_m\to\infty} v^i_{k_m}, \quad i = 1, \dots, s.$$

It is easy to see that $v^1, \dots, v^s$ are orthonormal vectors in $\widetilde{L}^+(p)$. It follows now from (12.12) that

$$\dim L^+(p) \ge s.$$

Taking into account that $\dim L^{-}(P_{k_m}) = n - s$, we obtain analogously that

$$\dim L^{-}(p) \geq n - s.$$

As

$$\dim L^{+}(p) + \dim L^{-}(p) = n,$$

we conclude that

$$\dim L^{+}(p) = s, \ \dim L^{-}(p) = n - s.$$

Hence the chosen number is unique, and $\widetilde{L}^{+}(p)$, $\widetilde{L}^{-}(p)$ are linear subspaces of $L^{+}(p)$, $L^{-}(p)$ having the same dimensions. It is evident now that

$$\widetilde{L}^{+}(p) = L^{+}(p), \ \widetilde{L}^{-}(p) = L^{-}(p). \qquad \square$$

Remarks.

1. It follows from the proof of Lemma 12.2 that the dimensions of $L^{+}(p)$, $L^{-}(p)$ are locally constant at points of a hyperbolic set.
2. It is easy to show using similar arguments that the projector R_p defined above is continuous in the following sense: if $p_k \in I \in H(C, \lambda)$, $p_k \underset{k\to\infty}{\longrightarrow} p \in I$,

   $$w_k \in T_{pk}M, \ w_k \underset{k\to\infty}{\longrightarrow} w \in T_pM$$

   then

   $$R_{p_k} w_k \underset{k\to\infty}{\longrightarrow} R_p w. \tag{12.13}$$

We are going now to introduce basic geometric objects associated with hyperbolic sets—stable and unstable manifolds.

Let $p \in I \in H(C, \lambda)$. Define the stable manifold and the unstable manifold of p:

$$W^s(p) = \{x \in M : d(f^k(x), f^k(p)) \underset{k\to+\infty}{\longrightarrow} 0\},$$
$$W^u(p) = \{x \in M : d(f^k(x), f^k(p)) \underset{k\to-\infty}{\longrightarrow} 0\}.$$

The following theorem (Theorem 12.1, usually called the Stable Manifold Theorem for hyperbolic sets) describes basic properties of stable and unstable manifolds. One can prove this theorem applying methods used to prove Theorem 4.1. We leave it to the reader to obtain the complete proof. Denote by $D_a(p)$ the open ball of radius $a > 0$ centered at $p \in M$ (with respect to metric d).

Theorem 12.1. *Suppose that $f \in C^r(M)$, $r \geq 1$, and that I is a hyperbolic set of f. Then there exists $\varepsilon_0 > 0$ having the following properties. If $p \in I$, dim $L^+(p) = l$, then:*

(1) *there exist immersions b^s, b^u of class C^r:*

$$b^s : \mathbf{R}^l \to M,\ b^s(0) = p, b^s(\mathbf{R}^l) = W^s(p),$$
$$b^u : \mathbf{R}^{n-l} \to M, b^u(0) = p,\ b^u(\mathbf{R}^{n-l}) = W^u(p);$$

(2) *for any $\varepsilon \in (0, \varepsilon_0)$ there exist smooth discs (of class C^r) $W^s_\varepsilon(p)$, $W^u_\varepsilon(p)$ being components of intersections $W^s(p) \cap D_\varepsilon(p)$, $W^u(p) \cap D_\varepsilon(p)$ containing p and such that:*

(a) $T_pW^s_\varepsilon(p) = L^+(p)$, $T_pW^u_\varepsilon(p) = L^-(p)$;

(b) *for $x \in D_\varepsilon(p)\backslash W^s_\varepsilon(p)$ there exists $k_1 > 0$ such that*

$$d(f^{k_1}(x),\ f^{k_1}(p)) \geq \varepsilon;$$

(c) *for $x \in D_\varepsilon(p)\backslash W^u_\varepsilon(p)$ there exists $k_2 < 0$ such that*

$$d(f^{k_2}(x),\ f^{k_2}(p)) \geq \varepsilon.$$

If p is a hyperbolic periodic point of f then $W^s(p)$, $W^u(p)$ coincide with stable and unstable manifolds of p introduced in Chapter 5. In this case $W^s_\varepsilon(p)$, $W^u_\varepsilon(p)$ are the corresponding local stable and local unstable manifolds.

Let Ω be the nonwandering set of diffeomorphism f and let P be the set of periodic points of f.

Smale introduced in [39] the following property of f.

AXIOM A:

(1) Ω is a hyperbolic set; (2) the set P is dense in Ω.

We are going to prove now the following result describing the structure of trajectories of a diffeomorphism that satisfies Axiom A.

Theorem 12.2 [The Spectral Decomposition Theorem]. *If a diffeomorphism f satisfies Axiom A then there exists a unique decomposition*

$$\Omega = \Omega_1 \cup \cdots \cup \Omega_m.$$

Here Ω_i are compact disjoint invariant sets, and each Ω_i contains a dense trajectory.

It follows from Lemma 12.2 that the subspaces $L^+(p)$ and $L^-(p)$ are continuous with respect to p. The local stable manifold and the local unstable manifold are tangent to $L^+(p)$, $L^-(p)$ at p, so it is easy to see that the following statement is true.

Lemma 12.3. *Let I be a hyperbolic set. Then for each $p \in I$ there exists a neighborhood U_p such that for any $x, y \in U_p \cap I$ the manifolds $W^u(x)$ and $W^s(y)$ have a point of transversal intersection.*

Let us prove the following result which we also need to prove Theorem 12.2.

Lemma 12.4. *Let p, q be hyperbolic periodic points of the diffeomorphism f and let x_1, x_2 be points of transversal intersection for $W^u(p), W^s(q)$ and for $W^s(p)$, $W^u(q)$, respectively. Then x_1, x_2 are nonwandering points of f.*

Proof. Suppose for definiteness that p, q are fixed points of f. To prove the general statement one can take f^k with large k instead of f.

Let V be an arbitrary neighborhood of x_1. Consider smooth closed discs Δ_s, Δ_u in $W^s(p)$, $W^u(q)$ respectively, such that x_2 is a point of transversal intersection of Int Δ_s, Int Δ_u (we take interiors with respect to the inner topologies of the discs). It follows from Theorem 6.3 that there exists $\delta > 0$ such that if

$$h_s \in E^1(\Delta_s, M),\ h_u \in E^1(\Delta_u, M), \text{ and } \rho_1(h_s, id) < \delta,$$

$\rho_1(h_u, id) < \delta$ then the discs $h_s(\Delta_s)$, $h_u(\Delta_u)$ have a point of transversal intersection.

The neighborhood V is transversal to $W^s(q)$. We obtain from the λ-lemma that there exists a disc $Q_u \subset W^u(q)$ and a sequence of embeddings $h_u^k \in E^1(Q_u, M)$ such that

$$\rho_1(h_u^k, id) \underset{k\to\infty}{\longrightarrow} 0,\ h_u^k(Q_u) \subset f^k(V).$$

The disc Q_u contains a neighborhood of q in $W^u(q)$, hence we can find m such that $f^m(\Delta_u) \subset Q_u$. The maps

$$\tilde{h}_u^k = f^{-m} \circ h_u^k|_{f^m(\Delta u)}$$

are embeddings of the disc Δ_u in M,

$$\rho_1(\tilde{h}_u^k, id) \underset{k\to\infty}{\longrightarrow} 0 \text{ in } E^1(\Delta_u, M),$$

and the discs $\Delta_u^k = \tilde{h}_u^k(\Delta_u)$ are subsets of $f^{k+m}(V)$.

Similarly we construct embeddings $\tilde{h}_s^k$ of the disc Δ_s in M such that

$$\rho_1(\tilde{h}_s^k, id) \underset{k\to\infty}{\longrightarrow} 0 \text{ in } E^1(\Delta_s, M).$$

We conclude that there exist $k_1 > 0$ and $k_2 < 0$ such that the discs $\Delta_u^{k_1}$ and $\Delta_s^{k_2} = \tilde{h}_s^{k_2}(\Delta_s)$ have a point of intersection. Hence, $f^{l_1}(V) \cap f^{l_2}(V) \neq \emptyset$ for some $l_1 < 0$, $l_2 > 0$. Therefore, $V \cap f^{l_2 - l_1}(V) \neq \emptyset$, and we can choose $l_2 - l_1$ arbitrarily large. We see that the point x_1 is nonwandering. One proves similarly that $x_2 \in \Omega$. □

Let us prove now Theorem 12.2. Let K be a set in M, and denote

$$T(K) = \bigcup_{x\in K,\, m\in\mathbb{Z}} f^m(x).$$

Take a point $x_0 \in \Omega$ and find a corresponding neighborhood U_0 having the property described in Lemma 12.3 (we take Ω as the hyperbolic set I).

Define the following set

$$\Omega_{x_0} = \overline{T(U_0 \cap \Omega)}.$$

The set Ω_{x_0} is the closure of an invariant subset of Ω; hence Ω_{x_0} is closed, invariant and $\Omega_{x_0} \subset \Omega$. Let us show that the set Ω_{x_0} depends only on x_0 and does not depend on the choice of U_0.

Lemma 12.5. *Let V be an open subset of U_0 such that $\Omega \cap V \neq \emptyset$. Then the set*

$$\Omega_V = \overline{T(V \cap \Omega)}$$

coincides with Ω_{x_0}.

Proof. The set P of periodic points is dense in Ω. Find a point $p \in P \cap V$. Let q be an arbitrary point in $P \cap U_0$. It follows from the choice of U_0 that there exist points x_1, x_2 of the transversal intersection of $W^u(p)$, $W^s(q)$ and of $W^s(p)$, $W^u(q)$, respectively. Lemma 12.4 shows that $x_1, x_2 \in \Omega$. As $x_1 \in W^u(p)$, the trajectory of x_1 intersects V, and hence $x_1 \in \Omega_V$. Since $x_1 \in W^s(q)$, any neighborhood of q contains points of the trajectory $f^k(x_1)$ and therefore, $q \in \overline{\Omega}_V$. The set Ω_V is closed; hence $q \in \Omega_V$. Consider now an arbitrary point $x \in U_0 \cap \Omega$. Any neighborhood of x contains a periodic point $q \in U_0$. We showed above that $q \in \Omega_V$; hence, $x \in \Omega_V$. This implies that

$$T(U_0 \cap \Omega) \subset \Omega_V.$$

As Ω_V is compact, we obtain that $\Omega_{x_0} \subset \Omega_V$. The inclusion $\Omega_V \subset \Omega_{x_0}$ is evident. □

Corollary. *For any* $\xi \in \Omega_{x_0}$, $\Omega_\xi = \Omega_{x_0}$.

Proof. Take $\xi \in \Omega_{x_0}$, and take an arbitrary neighborhood U_ξ of ξ. Define the set

$$\Omega_\xi = \overline{T(U_\xi \cap \Omega)}.$$

It follows from the definition of the set Ω_{x_0} that there exists $x_1 \in U_0 \cap \Omega$ and k_1 such that $f^{k_1}(x_1) \in U_\xi$. Then we can find an open set $V \subset U_0$ such that $x_1 \in V$ and $f^{k_1}(V) \subset U_\xi$. Lemma 12.5 shows that

$$\Omega_{x_0} = \overline{T(V \cap \Omega)} \subset \overline{T(U_\xi \cap \Omega)} = \Omega_\xi. \tag{12.15}$$

It follows from (12.15) that $x_0 \in \Omega_\xi$. Similar arguments prove that $\Omega_\xi \subset \Omega_{x_0}$. That completes the proof of the corollary. □

Let us construct sets Ω_x for all $x \in \Omega$. It follows from the Corollary of Lemma 12.5 that, for $x, y \in \Omega$, either $\Omega_x \cap \Omega_y = \emptyset$ or $\Omega_x = \Omega_y$. Indeed, if $\xi \in \Omega_x \cap \Omega_y$,

$$\Omega_x = \Omega_\xi = \Omega_y.$$

We claim that Ω is a finite union of sets Ω_x. To get a contradiction suppose that there exists a countable sequence $\Omega_{x_1}, \Omega_{x_2}, \dots, \Omega_{x_k}, \dots$ of disjoint sets Ω_x. Take points $x_k \in \Omega_{x_k}$ and consider a limit point x_0 of the sequence x_k. The points $x_k \in \Omega$, the set Ω is closed, and hence

$x_0 \in \Omega$. Consider the neighborhood U_{x_0} of x_0 used to construct the set Ω_{x_0}. There exists k_0 such that $x_k \in U_{x_0}$ for $k \geq k_0$. Therefore the sets Ω_{x_k}, $k \geq k_0$, coincide with Ω_{x_0}. The contradiction we obtained shows that Ω is a finite union of sets Ω_x. We denote these sets by $\Omega_1, \dots, \Omega_m$.

To prove that each set Ω_i, $i = 1, \dots, m$, contains a dense trajectory we apply a construction introduced by G. Birkhoff [17]. Consider a set Ω_i and consider a countable dense set p_k, $k = 1, 2, \dots$, in Ω_i. Take $\delta \in (0, 1)$. It is easy to see that any set $\tilde{p}_k$, $k = 1, 2, \dots$, such that $d(p_k, \tilde{p}_k) < \delta^k$, is also dense in Ω_1. We denote as above by $D_a(p)$ the ball of radius $a > 0$ centered at p, and let $D_k = D_{\delta^k}(p_k)$. The set D_1 is open and we can find a point $p_1 \in \Omega_i \cap D_1$. It follows from Lemma 12.5 and from the corollary of this lemma that

$$\Omega_i = \overline{T(D_1 \cap \Omega)}.$$

Consequently, there exist a point $p_2 \in D_1 \cap \Omega_i$ and a number k_2 such that

$$f^{k_2}(p_2) \in D_2.$$

Then there exists a neighborhood Q_2 of p_2 such that $\overline{Q}_2 \subset D_1$ and

$$f^{k_2}(\overline{Q}_2) \subset D_2.$$

Similar arguments show that there exists a point $p_3 \in Q_2 \cap \Omega_i$ and a number k_3 such that

$$f^{k_3}(p_3) \in D_3.$$

Find a neighborhood Q_3 of p_3, such that $\overline{Q}_3 \subset Q_2$ and

$$f^{k_3}(\overline{Q}_3) \subset D_3.$$

It follows from the construction that

$$f^{k_2}(\overline{Q}_3) \subset D_2.$$

Hence for any point $x \in \overline{Q}_3$ the trajectory $f^k(x)$ intersects D_2 and D_3. Continuing the process, we can find points p_j, numbers k_j, and neighborhoods Q_j of p_j such that

$$p_j \in Q_j \cap \Omega_i,\ f^{k_j}(\overline{Q}_j) \subset D_j,$$
$$D_1 \supset \overline{Q}_2 \supset Q_2 \supset \cdots \supset \bar{Q}_j \supset Q_j \supset \cdots \quad .$$

We require in addition that

$$\max_{x \in \overline{Q}_j} d(x, p_j) \underset{j \to \infty}{\longrightarrow} 0. \tag{12.16}$$

The intersection of the sequence $\overline{Q}_j$ of enclosed compacts is non-empty. It follows from (12.16) that this intersection is a point. Denote this point by $\tilde{p}$. We obtain from (12.16) that $p_j \to \tilde{p}$ as $j \to \infty$. As $p_j \in \Omega_i$, we see that $\tilde{p} \in \Omega_i$. We conclude from the description of the process that

$$f^{k_j}(\tilde{p}) \in D_j$$

and hence, the trajectory of the point $\tilde{p}$ is dense in Ω_i.

To complete the proof of Theorem 12.2 we are to show now that the decomposition (12.14) is unique. Suppose that

$$\Omega = \widetilde{\Omega}_1 \cup \cdots \cup \widetilde{\Omega}_{m'}$$

where $\widetilde{\Omega}_i$ are compact disjoint invariant sets having dense trajectories.

Consider a point $x \in \widetilde{\Omega}_1$ such that

$$\widetilde{\Omega}_1 = \overline{\bigcup_{k \in \mathbb{Z}} f^k(x)}.$$

The point $x \in \Omega$, and hence there exists i, $1 \leq i \leq m$, such that $x \in \Omega_i$. It was shown that in this case $\Omega_i = \Omega_x$, so we have

$$\widetilde{\Omega}_1 = \overline{\bigcup_{k \in \mathbb{Z}} f^k(x)} \subset \Omega_x = \Omega_i.$$

Let us show that $\Omega_i \subset \widetilde{\Omega}_1$. To get a contradiction suppose that there exists $\xi \in \Omega_i \backslash \widetilde{\Omega}_1$. Consider a point $p \in \Omega_i$ such that

$$\Omega_i = \overline{\bigcup_{k \in \mathbb{Z}} f^k(p)}$$

and choose sequences l_j, m_j such that

$$f^{l_j}(p) \to x, \ f^{m_j}(p) \to \xi \text{ as } j \to \infty.$$

The sets $\widetilde{\Omega}_1, \ldots, \widetilde{\Omega}_{m'}$ are disjoint compacts. Find their disjoint neighborhoods $U_1, \ldots, U_{m'}$. Take a neighborhood V of the set $\widetilde{\Omega}_1$ so small that

$$\overline{V \cup f(V)} \subset U_1, \quad \xi \notin \overline{V \cup f(V)},$$

If j is large enough then

$$f^{l_j}(p) \in V, \quad f^{m_j}(p) \notin \overline{V \cup f(V)},$$

and hence there exists ν_j such that

$$f^{\nu_j}(p) \in \overline{f(V)} \backslash V.$$

This is in contradiction with the inclusion $f^{\nu_j}(p) \in \Omega$. Similar arguments show that each set $\widetilde{\Omega}_j$, $j = 1, \ldots, m'$, coincides with one of the sets $\Omega_1, \ldots, \Omega_m$. □

The sets $\Omega_1, \ldots, \Omega_m$ in the decomposition (12.14) are called basic sets. For a compact $Q \subset M$ denote by $d(x, Q)$ the distance between a point x and Q. Define the following sets

$$W^s(\Omega_i) = \{x \in M : d(f^k(x), \Omega_i) \underset{k \to +\infty}{\longrightarrow} 0\},$$
$$W^u(\Omega_i) = \{x \in M : d(f^k(x), \Omega_i) \underset{k \to -\infty}{\longrightarrow} 0\},$$

$i = 1, \ldots, m$. These sets are analogues of stable and unstable manifolds for individual hyperbolic trajectories.

Theorem 12.3. *If a diffeomorphism f satisfies Axiom A then*

$$M = \bigcup_{1 \le i \le m} W^s(\Omega_i) = \bigcup_{1 \le i \le m} W^u(\Omega_i).$$

Proof. Consider a point $x \in M$. Arguments similar to the proof of Lemma 2.1 show that α_x, the α-limit set of the trajectory $f^k(x)$, and ω_x, the ω-limit set of $f^k(x)$, are subsets of Ω.

Let us show that there exists a basic set Ω_i such that $\omega_x \in \Omega_i$. It is evident that in this case $x \in W^s(\Omega_i)$. To get a contradiction suppose that there exist two basic sets Ω_i, Ω_j, $i \neq j$ such that

$$\omega_x \cap \Omega_i \neq \emptyset, \quad \omega_x \cap \Omega_j \neq \emptyset.$$

Take disjoint neighborhoods $U_1, \ldots, U_m$ of the basic sets $\Omega_1, \ldots, \Omega_m$. Find a neighborhood V of Ω_i such that

$$\overline{V \cup f(V)} \subset U_i.$$

There exist sequences $l_k, m_k \underset{k\to\infty}{\longrightarrow} +\infty$ such that $l_k < m_k < l_{k+1}$ and

$$d(f^{l_k}(x), \Omega_i) \underset{k\to\infty}{\longrightarrow} 0, \; d(f^{m_k}(x), \Omega_j) \underset{k\to\infty}{\longrightarrow} 0.$$

If k is large enough then $f^{l_k}(x) \in V$, $f^{m_k}(x) \in U_j$. Therefore we can find a sequence ν_k such that $l_k < \nu_k < m_k$ and

$$f^{\nu_k}(x) \in \overline{f(V)} \backslash V.$$

Consider a limit point ξ of the sequence $f^{\nu_k}(x)$. It is evident that $\xi \in \omega_x \subset \Omega$. On the other hand, ξ belongs to the compact $\overline{f(V)} \backslash V$, and $(\overline{f(V)} \backslash V) \cap \Omega = \emptyset$. This contradiction shows that ω_x is a subset of a unique Ω_i. Similar arguments show that α_x is a subset of a unique Ω_i. □

It follows from Theorem 12.3 that each trajectory $f^k(x)$, $x \in M$, tends to a basic set as $k \to +\infty$ and as $k \to -\infty$. The following more exact statement is also true.

Theorem 12.4. [7]. *If a diffeomorphism f satisfies Axiom A, then*

$$M = \bigcup_{p\in\Omega} W^s(p) = \bigcup_{p\in\Omega} W^u(p).$$

We prove Theorem 12.4 in Section 3 of this chapter after the proof of Lemma 12.9. We prove Theorem 12.4 under additional assumptions. That is, we consider diffeomorphisms having no 1-cycles (see definitions below).

Let us now give definitions that we need to formulate conditions for Ω-stability and for the structural stability of diffeomorphisms.

For two distinct basic sets Ω_i, Ω_j we write $\Omega_i \to \Omega_j$ if

$$W^u(\Omega_i) \cap W^s(\Omega_j) \neq \emptyset.$$

We say that f has a 1-cycle if there exists a basic set Ω_i such that

$$(W^u(\Omega_i) \cap W^s(\Omega_i)) \backslash \Omega_i \neq \emptyset.$$

We say that f has a k-cycle, $k > 1$, if there exist k different basic sets $\Omega_{i_1}, \ldots, \Omega_{i_k}$ such that

$$\Omega_{i_1} \to \Omega_{i_2} \to \cdots \to \Omega_{i_k} \to \Omega_{i_1}.$$

We say that f satisfies the no-cycle condition if f has no k-cycles with $k \geq 1$.

Theorem 12.5. (S. Smale, [40]). *If a diffeomorphism f satisfies Axiom A and the no-cycle condition then f is Ω-stable.*

We say that f satisfies the geometric strong transversality condition if for any $x, y \in \Omega$ the manifolds $W^u(x)$ and $W^s(y)$ are transversal. There is also the so called analytic strong transversality condition; we discuss the connection between these strong transversality conditions in Chapter 13.

J. Robbin (for diffeomorphisms of class C^2 [28]), and C. Robinson (for diffeomorphisms of class C^1 [30]) proved the following result.

Theorem 12.6. *If a diffeomorphism f satisfies Axiom A and the geometric strong transversality condition then f is stucturally stable.*

We prove Theorem 12.5 in Section 3 of this chapter. Proofs of Theorem 12.6 are too complicated to be presented here.

In a recent paper [13] R. Mañé proved that if a diffeomorphism f of class C^1 is structurally stable then it satisfies Axiom A and the geometric strong transversality condition. So the conditions of Theorem 12.6 are necessary and sufficient for the structural stability. Using techniques introduced by R. Mañé, J. Palis proved in [23] that the conditions of Theorem 12.5 are necessary and sufficient for Ω-stability.

2. We are going to prove now the structural stability of a hyperbolic set (Theorem 12.8). To simplify the proof we consider a hyperbolic set I of a diffeomorphism $f : \mathbb{R}^n \to \mathbb{R}^n$ of class C^1. We suppose that $I \in H(C, \lambda)$ and that N is chosen so that (12.7) holds.

Take a point $p \in I$ and denote $p_k = f^k(p)$ for $k \in \mathbb{Z}$. It follows from the Taylor formula that we can write

$$f(p + x) = f(p) + Df(p)x + g(p, x), \tag{12.17}$$

where $g(p,0)=0$ and

$$\sup_{\substack{p\in I; x_1\neq x_2\\ |x_1|,|x_2|\leq a}} \frac{|g(p,x_1)-g(p,x_2)|}{|x_1-x_2|} \underset{a\to 0}{\longrightarrow} 0. \tag{12.18}$$

Consider two integers k, m such that $m \leq k-1$, and define the following map $\mathbb{R}^n \to \mathbb{R}^n$:

$$v \mapsto Df^k(p)R_p Df^{-m}(p_m)v.$$

Here R_p is the projector onto $L^+(p)$ defined before Lemma 12.1.

Let us prove that

$$\|Df^k(p)R_pDf^{-m}(p_m)\| \leq C_1\lambda^{k-m} \tag{12.19}$$

where $C_1 = C\alpha_1$ (α_1 is introduced in the Corollary of Lemma 12.2). Let us represent

$$v = v^+ + v^-,\ v^+ \in L^+(p_m),\ v^- \in L^-(p_m).$$

Then

$$Df^{-m}(p_m)v = Df^{-m}(p_m)v^+ + Df^{-m}(p_m)v^-.$$

We obtain from (12.2)

$$Df^{-m}(p_m)v^+ \in L^+(p),\ \ Df^{-m}(p_m)v^- \in L^-(p),$$

so

$$R_pDf^{-m}(p_m)v = Df^{-m}(p_m)v^+.$$

Then

$$\begin{aligned}|Df^k(p)R_pDf^{-m}(p_m)v| &= |Df^{k-m}(p)v^+| \\ &\leq C\lambda^{k-m}|v^+| \leq C\alpha_1\lambda^{k-m}|v|.\end{aligned}$$

We take into account here that $v^+ = R_{p_m}v$, and use the inequality (12.10). So, (12.19) is established.

Similar arguments show that if k, m are integers with $m > k$ then

$$\|Df^k(p)(E_p - R_p)Df^{-m-1}(p_{m+1})\| \leq C_2\lambda^{m-k}, \tag{12.20}$$

where $C_2 = C_1N$.

We investigate now the problem of the existence of trajectories of a diffeomorphism $\tilde{f}$, C^1-close to f, which are close to trajectories of the set I.

Fix a neighborhood V of I such that $\overline{V}$ is compact. Define the following distance between diffeomorphisms $f, \tilde{f}$:

$$\rho_1(f,\tilde{f}) = \sup_{x\in\overline{V}} \left(|f(x) - \tilde{f}(x)| + \|\frac{\partial f}{\partial x}(x) - \frac{\partial \tilde{f}}{\partial x}(x)\| \right).$$

We are going to use results we obtain in Section 3, studying the diffeomorphism of a smooth closed manifold M. It is easy to see that if two diffeomorphisms $f, \tilde{f}$ of M are C^1-close in $\mathrm{Diff}^1(M)$, then they are also C^1-close with respect to the introduced metric ρ_1.

Fix $p \in I$ and define the map $\tilde{f}$ by

$$\tilde{f}(p+x) = f(p) + Df(p)x + h(p,x) \tag{12.21}$$

for $x \in \mathbf{R}^n$. We suppose that the function h in (12.21) satisfies the following conditions:

$$|h(p,0)| \le l_1, \tag{12.22}$$

$$|h(p,x_1) - h(p,x_2)| \le l_2|x_1 - x_2|. \tag{12.23}$$

Consider the Banach space $l^\infty(\mathbf{Z})$ of bounded sequences $\eta = \{\eta_k\}$, $k \in \mathbf{Z}$, with the following norm

$$|\eta|^\infty = \sup_{k\in\mathbf{Z}} |\eta_k|.$$

Define now the operator $\Xi(p,h)$ corresponding to the trajectory $p_m = f^m(p)$ and to the map (12.21) such that Ξ maps a sequence $\eta = \{\eta_k\}$ on the sequence $\xi = \{\xi_k\}$ in the following way:

$$\xi_k = \sum_{m=-\infty}^{k} Df^k(p)R_p Df^{-m}(p_m)h(p_{m-1},\eta_{m-1})$$
$$- \sum_{m=k}^{\infty} Df^k(p)(E_p - R_p)Df^{-m-1}(p_{m+1})h(p_m,\eta_m). \tag{12.24}$$

Let us denote by $h(p,\eta)$ the sequence $\{h(p_m,\eta_m)\}$. It follows from (12.22)and (12.23) that

$$|h(p,0)|^\infty \le l_1, \; |h(p,\eta)|^\infty \le l_1 + l_2|\eta|^\infty. \tag{12.25}$$

Lemma 12.6. *There exists $\beta = \beta(N, C, \lambda)$ such that for $\xi, \eta \in l^\infty(\mathbf{Z})$ we have*

$$|\Xi(p,h)(\eta)|^\infty \le \beta(l_1 + l_2|\eta|^\infty), \tag{12.26}$$

$$|\Xi(p,h)(\xi) - \Xi(p,h)(\eta)|^\infty \le \beta l_2 |\xi - \eta|^\infty. \tag{12.27}$$

Proof. Let us estimate

$$\begin{aligned} |\xi_k| &\le \sum_{m=-\infty}^{k} \|Df^k(p)R_p Df^{-m}(p_m)\| \, |h(p,\eta)|^\infty \\ &+ \sum_{m=k}^{\infty} \|Df^k(p)(E_p - R_p)Df^{-m-1}(p_{m+1})\| \, |h(p,\eta)|^\infty \\ &\le \left(\sum_{m=-\infty}^{k} C_1 \lambda^{k-m} + \sum_{m=k}^{\infty} C_2 \lambda^{m-k} \right) |h(p,\eta)|^\infty \\ &= [C_1(1 + \lambda + \lambda^2 + \cdots) + C_2(1 + \lambda + \cdots)] |h(p,\eta)|^\infty \\ &\le \beta(l_1 + l_2|\eta|^\infty), \end{aligned}$$

where

$$\beta = \frac{C_1 + C_2}{1 - \lambda}.$$

So, (12.26) is established. Similar arguments prove (12.27). □

To describe the connection of the operator Ξ with trajectories of the map (12.21) we need the following lemma.

Lemma 12.7. *Let $\eta = \{\eta_k\} \in l^\infty(\mathbf{Z})$. The equalities $\tilde{f}(p_k + \eta_k) = p_{k+1} + \eta_{k+1}$, $k \in \mathbf{Z}$, hold if and only if*

$$\Xi(p,h)(\eta) = \eta. \tag{12.28}$$

Proof. Denote $\xi = \Xi(p,h)(\eta)$. If (12.28) is true then

$$\tilde{f}(p_k + \eta_k) = f(p_k) + Df(p_k)\eta_k + h(p_k, \eta_k)$$

$$= p_{k+1} + Df(p_k) \sum_{m=-\infty}^{k} Df^k(p) R_p Df^{-m}(p_m) h(p_{m-1}, \eta_{m-1})$$
$$- Df(p_k) \sum_{m=k}^{\infty} Df^k(p)(E_p - R_p) Df^{-m-1}(p_{m+1}) h(p_m, \eta_m)$$
$$+ h(p_k, \eta_k). \tag{12.29}$$

As $Df(p_k)Df^k(p) = Df^{k+1}(p)$, the second term in the right side of (12.29) equals

$$\sum_{m=-\infty}^{k+1} Df^{k+1}(p) R_p Df^{-m}(p_m) h(p_{m-1}, \eta_{m-1})$$
$$-Df^{k+1}(p) R_p Df^{-k-1}(p_{k+1}) h(p_k, \eta_k),$$

and the third term equals

$$- \sum_{m=k+1}^{\infty} Df^{k+1}(p)(E_p - R_p) Df^{-m-1}(p_{m+1}) h(p_m, \eta_m)$$
$$-Df^{k+1}(p)(E_p - R_p) Df^{-k-1}(p_{k+1}) h(p_k, \eta_k).$$

As

$$Df^{k+1}(p) R_p Df^{-k-1}(p_{k+1}) + Df^{k+1}(p)(E_p - R_p) Df^{-k-1}(p_{k+1})$$
$$= Df^{k+1}(p) E_p Df^{-k-1}(p_{k+1}) = E_p$$

the sum of the second term and of the third term in (12.29) equals $\xi_{k+1} - h(p_k, \eta_k)$. Hence, the equality $\tilde{f}(p_k + \eta_k) = p_{k+1} + \eta_{k+1}$ is equivalent to the equality $\xi_{k+1} = \eta_{k+1}$. The proof of the inverse statement is similar. □

Let us show now that if a diffeomorphism $\tilde{f}$ is C^1-close to f and if a part of a trajectory of $\tilde{f}$ belongs to a small neighborhood of I then this part of a trajectory is also "hyperbolic". To prove this statement (Theorem 12.7) we apply the method described in [24] and using the Stable Manifold Theorem.

Theorem 12.7. *Let I be a hyperbolic set for diffeomorphism f, $I \in H(C, \lambda)$, and let ε_0 be the number introduced for I in Theorem*

12.1. Then given $\delta > 0$ there exists a neighborhood U_δ of I and $\Delta > 0$ such that

(1) *if for a point $p \in U_\delta$ and for a diffeomorphism $\tilde{f}$ with $\rho_1(f, \tilde{f}) < \Delta$ the points*

$$p_k = \tilde{f}^k(p) \in U_\delta \text{ for } k_1 < k < k_2$$

(we allow $k_1 = -\infty$ and $k_2 = +\infty$) then for each p_k there exist linear subspaces $\widetilde{L}^+(p_k)$, $\widetilde{L}^-(p_k)$ having the following properties:

(a) $\widetilde{L}^+(p_k) + \widetilde{L}^-(p_k) = \mathbf{R}^n$;

(b) $D\tilde{f}(p_k)\widetilde{L}^+(p_k) = \widetilde{L}^+(p_{k+1})$,
$D\tilde{f}(p_k)\widetilde{L}^-(p_k) = \widetilde{L}^-(p_{k+1})$, $k_1 < k < k_2 - 1$;

(c) *for $v \in \widetilde{L}^+(p_k)$, $0 \leq m < k_2 - k$ we have*

$$|D\tilde{f}^m(p_k)v| \leq (C + \delta)(\lambda + \delta)^m |v|; \qquad (12.30)$$

(d) *For $v \in \widetilde{L}^-(p_k)$, $k_1 - k < m \leq 0$ we have*

$$|D\tilde{f}^m(p_k)v| \leq (C + \delta)(\lambda + \delta)^{-m} |v|; \qquad (12.31)$$

(2) *if for $p \in U_\delta$ and for $\tilde{f}$ with $\rho_1(f, \tilde{f}) < \Delta$*

$$\tilde{f}^k(p) \in U_\delta \text{ for } k \geq 0$$

then for any $\varepsilon \in (0, \varepsilon_0 - \delta)$ there exists a smooth disc $\widetilde{W}^s_\varepsilon(p)$ containing p and such that

$$T_p\widetilde{W}^s_\varepsilon(p) = \widetilde{L}^+(p),$$

for $y \in \widetilde{W}^s_\varepsilon(p)$, $k \geq 0$ the inequality

$$|\tilde{f}^k(y) - \tilde{f}^k(p)| \leq (C + \delta)(\lambda + \delta)^k |y - p| \qquad (12.32)$$

holds, and if $y \in D_\varepsilon(p) \backslash \widetilde{W}^s_\varepsilon(p)$ then there exists $\kappa > 0$ such that

$$|\tilde{f}^\kappa(y) - \tilde{f}^\kappa(p)| \geq \varepsilon;$$

(3) *if for $p \in U_\delta$ and for $\tilde{f}$ with $\rho_1(f, \tilde{f}) < \Delta$*

$$\tilde{f}^k(p) \in U_\delta \text{ for } k \leq 0$$

then for any $\varepsilon \in (0, \varepsilon_0 - \delta)$ *there exists a smooth disc* $\widetilde{W}^u_\varepsilon(p)$ *having properties analogous to the properties of* $\widetilde{W}^s_\varepsilon(p)$ *(changing* s *by* u, $k \geq 0$ *by* $k \leq 0$ *etc).*

Proof. Fix a neighborhood V of I such that $\overline{V}$ is compact. Take $N_0 > 0$ such that

$$\|Df(x)\| \leq N_0 \text{ for } x \in \overline{V}.$$

Consider arbitrary $\delta_0 > 0$. It follows from Lemma 12.2 and from the uniform continuity of $Df(x)$ on $\overline{V}$ that there exists $\delta_1 > 0$ having the following properties:

(1) for $x, y \in I$ with $|x - y| < \delta_1$ one can find a nonsingular $n \times n$ matrix $Q(x, y)$ such that

$$Q(x,y)L^+(x) = L^+(y), Q(x,y)L^-(x) = L^-(y),$$
$$|Q(x,y)v| = |v| \text{ for } v \in L^+(x) \cup L^-(x),$$
$$\|Q(x,y) - E\| \leq \delta_0;$$

(2) if $x, y \in \overline{V}$ and $|x - y| < \delta_1$, then

$$\|Df(x) - Df(y)\| \leq \delta_0 \tag{12.33}$$

Find $\delta_2 > 0$ such that for $x, y \in I$ the inequality $|x - y| < \delta_2$ implies

$$|f(x) - f(y)| < \frac{\delta_1}{3}. \tag{12.34}$$

We suppose that $3\delta_2 < \delta_1 < \delta_0$. Denote by U^* the δ_2-neighborhood of I. Suppose that $U^* \subset V$. Consider a diffeomorphism $\tilde{f}$ such that $\rho_1(f, \tilde{f}) < \frac{\delta_1}{3}$.

Consider a point $p \in U^*$ such that

$$p_k = \tilde{f}^k(p) \in U^* \text{ for } k_1 < k < k_2.$$

Find for each p_k a point $\xi_k \in I$ such that $|p_k - \xi_k| < \delta_2$.

Define for p_k two linear subspaces of $\mathbb{R}^n$:

$$\widehat{L}^+(p_k) = L^+(\xi_k), \widehat{L}^-(p_k) = L^-(\xi_k)$$

and the linear map

$$\widehat{D}(p_k) = Q(f(\xi_k), \xi_{k+1})Df(\xi_k).$$

It follows from (12.34) that

$$|f(\xi_k) - f(p_k)| < \frac{\delta_1}{3}.$$

As $\tilde{f}(p_k) = p_{k+1}$,

$$|\tilde{f}(p_k) - f(p_k)| < \frac{\delta_1}{3}, \ |\xi_{k+1} - p_{k+1}| < \frac{\delta_1}{3},$$

we obtain that

$$|f(\xi_k) - \xi_{k+1}| < \delta_1.$$

Hence, the matrix $Q(f(\xi_k), \xi_{k+1})$ is well-defined and possesses the properties described above. It is easy to see that

$$\widehat{D}(p_k)\widehat{L}^+(p_k) = Q(f(\xi_k), \xi_{k+1})L^+(f(\xi_k)) = L^+(\xi_{k+1}) = \widehat{L}^+(p_{k+1}).$$

It follows from the properties of $Q, L^+, \widehat{L}^+$ that for $v \in \widehat{L}^+(p_k)$, $m > 0$ (such that $m + k < k_2$) we have

$$|\widehat{D}(p_{k+m+1}) \cdots \widehat{D}(p_{k+1})\widehat{D}(p_k)v| \leq C\lambda^m |v|. \tag{12.35}$$

Similarly for $v \in \widehat{L}^-(p_k)$, $m < 0$ (such that $m + k > k_1$)

$$|\widehat{D}^{-1}(p_{k+m+1}) \cdots \widehat{D}^{-1}(p_{k-1})\widehat{D}^{-1}(p_k)v| \leq C\lambda^{-m}|v|. \tag{12.36}$$

Let us estimate

$$\begin{aligned}\|D\tilde{f}(p_k) - \widehat{D}(p_k)\| &\leq \|Q(f(\xi_k), \xi_{k+1})(Df(\xi_k) - Df(p_k)\| \\ &\quad + \|Q(f(\xi_k), \xi_{k+1})Df(p_k) - Df(p_k)\| + \|Df(p_k) - D\tilde{f}(p_k)\| \\ &\leq (1 + \delta_0)\delta_0 + N_0\delta_0 + \delta_0 = \delta_0(N_0 + 2 + \delta_0).\end{aligned}$$

We take into account that $\|Q\| < 1 + \delta_0$ and that (12.33) holds. Consider the map

$$\varphi(p_k + x) = p_{k+1} + D\tilde{f}(p_k)x.$$

Evidently $\varphi(p_k) = p_{k+1}$. It was shown above that we can write

$$\varphi(p_k + x) = p_{k+1} + \widehat{D}(p_k)x + g_k x \tag{12.37}$$

where the term $g_k x$ has Lipschitz constant equal to $\delta_0(N_0 + 2 + \delta_0)$.

Let us apply to the map (12.37) Perron's method for the constructing of stable and unstable manifolds. We described the method in the case of a hyperbolic rest point of an autonomous system of differential equations in the proof of Theorem 4.1. Analyzing the proof of Theorem 4.1, one can see that we use actual inequalities (4.6)and (4.7) being analogous to inequalities (12.35)and (12.36) and estimates of Lipschitz constants of F_1, F_2. In estimates (4.10), (4.11) and analagous estimates we take constants $\sigma = \frac{\lambda}{2}, 2a$. It is easy to see that we can take any $\tilde{\sigma} \in (0, \lambda)$ instead of σ, and any $\tilde{a} > a$ instead of $2a$ (reducing respectively Lipschitz constants of the nonlinearities).

So, given $\delta > 0$ we can choose δ_0 so small that the map φ has the following property: for each point p_k there exist discs $\widetilde{L}^+(p_k)$, $\widetilde{L}^-(p_k)$ such that

$$\dim \widetilde{L}^+(p_k) = \dim L^+(\xi_k), \dim \widetilde{L}^-(p_k) = \dim L^-(\xi_k),$$

and for $v \in \widetilde{L}^+(p_k)$ with small $|v|$, (12.30) holds, and for $v \in \widetilde{L}^-(p_k)$ with small $|v|$, (12.31) holds. As the map φ is linear with resepct to x, and the multiplier of x coincides with $Df(p_k)$, it is easy to see that discs $\widetilde{L}^+(p_k)$, $\widetilde{L}^-(p_k)$ are parts of linear subspaces of $\mathbf{R}^n$. These spaces are invariant with respect to $D\tilde{f}(p_k)$, and (12.30), (12.31) hold for any vector v of the corresponding subspace. Thus we have proved the first statement of the theorem.

To prove the second statement use the first one and repeat the arguments of Perron's method (taking into account previous remarks about constants). □

Let us prove now the following statement playing a crucial role in proving the structural stability of a hyperbolic set.

Lemma 12.8. *Let I be a hyperbolic set of diffeomorphism f. There exists a neighborhood U_0 of the set I, a neighborhood W_0 of f (with respect to metric ρ_1) and $\sigma > 0$ such that given $\varepsilon > 0$ there exists $\delta > 0$ having the following property: if $f_1, f_2 \in W_0$ and $\rho_1(f_1, f_2) < \delta$ then for any point p_1 such that $f_1^k(p_1) \in U_0$, $k \in \mathbf{Z}$, there exists a point p_2 with*

$$|f_1^k(p_1) - f_2^k(p_2)| < \varepsilon, \ k \in \mathbf{Z},$$

and for any point $p' \neq p_2$ there exists $k \in \mathbf{Z}$ such that

$$|f_1^k(p_1) - f_2^k(p') \geq \sigma.$$

Proof. Let $I \in H(C, \lambda)$. Take $\delta_0 > 0$ such that $\lambda + \delta_0 < 1$. Apply Theorem 12.7 and find $\Delta > 0$ and a neighborhood U_{δ_0} corresponding to δ_0 (we denote this neighborhood as U_0). Denote by W_1 the set of diffeomorphisms $\tilde{f}$ such that $\rho_1(f, \tilde{f}) < \Delta$. It follows from Theorem 12.7 that any trajectory

$$\tilde{f}^k(p) \in U_0, \ k \in \mathbf{Z}, \tag{12.38}$$

for $\tilde{f} \in W_1$ is hyperbolic with constants $\widetilde{C} \leq C + \delta$, $\tilde{\lambda} \leq \lambda + \delta_0$ (we do not give exact definitions for these evident notions). Obviously

$$|D\tilde{f}(x)| \leq \widetilde{N} = N_0 + \delta, \ x \in U_0,$$

for $\tilde{f} \in W_1$. Therefore we can find the same numbers α, α_1 (see Lemma 12.1 and its corollary) for any $\tilde{f} \in W_1$. For any point p such that (12.38) holds, and for any map $h(p, x)$, we can define the map

$$\widetilde{\Xi}(p, h) : l^\infty(\mathbf{Z}) \to l^\infty(\mathbf{Z})$$

associated with $\tilde{f}, p, h$ by a formula similar to (12.24). It is easy to see now that we can choose a number β such that (12.26), (12.27) hold for $\widetilde{\Xi}$ being the same for any $\tilde{f} \in W_1$.

Consider the presentation (12.17) for f. It follows from (12.18) that the Lipschitz constant of g with repect to x in the ball $|x| < a$ tends to zero as $a \to 0$. Take $\gamma > 0$ such that

$$\gamma\beta < \frac{1}{2} \tag{12.39}$$

and find $\sigma > 0$ such that

$$|g(p, x_1) - g(p, x_2)| \leq \frac{\gamma}{2}|x_1 - x_2|$$

for $p \in \overline{U}_0; |x_1|, |x_2| \leq \sigma$. Consider the presentation

$$\tilde{f}(p + x) = \tilde{f}(p) + D\tilde{f}(p)x + \tilde{g}(p, x) \tag{12.40}$$

of $\tilde{f}$ similar to (12.17). Evidently

$$\begin{aligned} \tilde{g}(p, x) = f(p) - \tilde{f}(p) + (Df(p) - D\tilde{f}(p))x \\ + g(p, x) + \Delta f(p, x), \end{aligned} \tag{12.41}$$

where

$$\Delta f(p,x) = \tilde{f}(p+x) - f(p+x).$$

We obtain from the inequalities

$$\|Df(p) - D\tilde{f}(p)\| \le \rho_1(f,\tilde{f}),\ \|\frac{\partial}{\partial x}(f(p+x) - \tilde{f}(p+x))\| \le \rho_1(f,\tilde{f})$$

and from (12.41) that there exists a neighborhood W_2 of f such that $W_2 \subset W_1$ and for any $\tilde{f} \in W_2$, $p \in \overline{U}_0$ the function $\tilde{g}(p,x)$ in (12.40) satisfies the inequality

$$|\tilde{g}(p,x_1) - \tilde{g}(p,x_2)| \le \frac{2\gamma}{3}|x_1 - x_2|,\ |x_1|,|x_2| \le \sigma. \tag{12.42}$$

Find $\delta_1 > 0$ such that

$$\frac{2\gamma}{3} + \delta_1 < \gamma \tag{12.43}$$

and denote by W_0 a neighborhood of f such that $W_0 \subset W_2$ and for $f_1, f_2 \in W_0$ we have

$$\rho_1(f_1,f_2) < \delta_1.$$

Consider diffeomorphisms $f_1, f_2 \in W_0$ and a point p_1 such that

$$f_1^k(p_1) \in U_0,\ \ k \in \mathbf{Z}.$$

Let $g_{12}(p) = f_2(p) - f_1(p)$, then

$$f_2(p+x) = f_1(p) + Df_1(p)x + g_1(p,x) + g_{12}(p+x).$$

Here g_1 plays the same role as g in (12.17). Denote

$$g^*(p,x) = g_1(p,x) + g_{12}(p+x)$$

It is easy to see that $|g^*(p,0)| \le \rho_1(f_1,f_2)$. It follows from (12.42), (12.43) and from the choice of W_0 that

$$|g^*(p,x_1) - g^*(p,x_2)| \le \gamma|x_1 - x_2|$$

for $p \in U_0; |x_1|,|x_2| \le \sigma$. Consider the operator

$$\Xi_1(p_1,g^*) : l^\infty(\mathbf{Z}) \to l^\infty(\mathbf{Z})$$

corresponding to the diffeomorphism f_1, to the point p_1, and to the function g^*. It follows from our previous considerations that for $\xi, \eta \in l^\infty(\mathbf{Z})$ with $|\xi|^\infty$, $|\eta|^\infty \leq \sigma$ we have

$$|\Xi_1(p_1, g^*)(\eta)|^\infty \leq \beta(\rho_1(f_1, f_2) + \gamma|\eta|^\infty), \tag{12.44}$$

$$|\Xi_1(p_1, g^*)(\xi) - \Xi_1(p_1, g^*)(\eta)|^\infty \leq \beta\gamma|\xi - \eta|^\infty. \tag{12.45}$$

Applying Lemma 12.7 we see that the sequence $\{f_1^k(p_1) + \eta_k\}$ is a trajectory of f_2 if and only if $\eta = \{\eta_k\}$ is a fixed point of $\Xi_1(p_1, g^*)$.

Given $\varepsilon > 0$ choose $\delta = \delta(\varepsilon)$ such that

$$\beta(\delta(\varepsilon) + \gamma\varepsilon) < \varepsilon. \tag{12.46}$$

Note that (12.39) implies the possibility of the choice of δ. Note also that δ depends only on ε and on the choice of U_0, W_0, and does not depend on the choice of f_1, f_2.

It follows from (12.44)–(12.46) that if $\rho_1(f_1, f_2) < \delta(\varepsilon)$ then the operator $\Xi_1(p_1, g^*)$ maps the ball $|\eta|^\infty < \varepsilon$ into itself being a contraction on this ball. Hence, there exists a unique fixed point $\tilde{\eta} = \{\tilde{\eta}_k\}$ with $|\tilde{\eta}|^\infty < \varepsilon$. The inequalities

$$|f_2^k|(p_2) - f_1^k(p_1)| < \varepsilon, \quad k \in \mathbf{Z},$$

evidently hold for $p_2 = p_1 + \tilde{\eta}_0$. We obtain from (12.45) that the operator $\Xi_1(p_1, g^*)$ is a contraction on the ball $|\eta|^\infty \leq \sigma$. Therefore, if $\varepsilon < \sigma$ then $\tilde{\eta}$ is the unique fixed point in the ball $|\eta|^\infty \leq \sigma$. So for any $p' \neq p_2$ there exists $k \in \mathbf{Z}$ such that

$$|f_1^k(p_1) - f_2^k(p')| \geq \sigma.$$

□

Theorem 12.8. *Let I be a hyperbolic set of a diffeomorphism f. Given $\varepsilon > 0$ there exists a neighborhood $W(\varepsilon)$ of f (with respect to metric ρ_1) such that for any $\tilde{f} \in W(\varepsilon)$ there exists a hyperbolic set $\tilde{I}$ of $\tilde{f}$ and a homeomorphism $h : I \to \tilde{I}$ conjugating f on I and $\tilde{f}$ on $\tilde{I}$ and such that*

$$|x - h(x)| < \varepsilon \ \text{ for } x \in I. \tag{12.47}$$

Proof. Let $I \in H(C, \lambda)$. Choose $\delta_0 > 0$ such that $\lambda + \delta_0 < 1$. Find the corresponding neighborhoods $U\delta_0$ of I and $W\delta_0$ of f described in Theorem 12.7. Find also neighborhoods U_0, W_0 and σ described in Lemma 12.8. We suppose that $U_0 \subset U_{\delta_0}$, $W_0 \subset W_{\delta_0}$.

Take arbitrary $\varepsilon > 0$ and suppose that $\varepsilon < \sigma$ and that the ε-neighborhood U_ε of I is a subset of U_0. It follows from Lemma 12.8 that there exists a neighborhood $W(\varepsilon)$ of f having the following properties: $W(\varepsilon) \subset W_0$ and for any $\tilde{f} \in W(\varepsilon)$, $p \in I$ there exists a point $\tilde{p}$ such that

$$|f^k(p) - \tilde{f}^k(\tilde{p})| < \varepsilon, \quad k \in \mathbf{Z}, \tag{12.48}$$

and for any $p' \neq \tilde{p}$

$$\max_{k \in \mathbf{Z}} |f^k(p) - \tilde{f}^k(p')| \geq \sigma. \tag{12.49}$$

We claim that $W(\varepsilon)$ has the desired properties. Take $\tilde{f} \in W(\varepsilon)$ and define the map $h : I \to \mathbb{R}^n$ in the following way: for $p \in I$ let $h(p) = \tilde{p}$. Here $\tilde{p}$ is the point for which (12.48) is satisfied. It follows from (12.49) and from the choice of ε that the map h is well-defined. If we let $k = 0$ in (12.48) we see that (12.47) holds. Let us show that for $p \in I$

$$h(f(p)) = \tilde{f}(h(p)).$$

Let $p_1 = f(p), \tilde{p}_1 = \tilde{f}(\tilde{p})$. As

$$|f^k(p_1) - \tilde{f}^k(\tilde{p}_1)| = |f^{k+1}(p) - \tilde{f}^{k+1}(\tilde{p})| < \varepsilon,$$

for $k \in \mathbf{Z}$, we obtain that $h(p_1) = \tilde{p}_1$ so that h conjugates f and $\tilde{f}$.

Let us now show that h is continuous. Consider $p_m \in I$ such that $p_m \to p_0 \in I$ as $m \to \infty$. Consider a limit point r of the sequence $h(p_m)$. Such a limit point exists as $h(I) \subset U_\varepsilon$. If $r \neq h(p_0)$ there exists k_0 such that

$$|f^{k_0}(p_0) - \tilde{f}^{k_0}(r)| \geq \sigma. \tag{12.50}$$

Taking into account that

$$|f^{k_0}(p_m) - \tilde{f}^{k_0}(h(p_m))| < \varepsilon$$

for all m and passing to the limit in the last inequality with respect to the sequence of indices for which $h(p_m) \to r$, we obtain that

$$|f^{k_0}(p_0) - \tilde{f}^{k_0}(r)| \leq \varepsilon.$$

The last inequality contradicts (12.50) and the choice of ε.

So, if we let $\widetilde{I} = h(I)$, h is a homeomorphism of I onto $\widetilde{I}$ satisfying every requirement of the theorem. As $\widetilde{I} \subset U_{\delta_0}$, we obtain from Theorem 12.7 that $\widetilde{I}$ is a hyperbolic set of $\tilde{f}$. □

3. Consider a diffeomorphism f of a smooth closed manifold M. We apply to hyperbolic sets of I all results on hyperbolic sets obtained in Section 2 (it is easy to see that all proofs of Section 2 are valid in the case of M requiring only the proper choices of coordinates).

We mentioned in Section 1 that f is called an Anosov diffeomorphism if M is a hyperbolic set of f.

Theorem 12.9. *An Anosov diffeomorphism f is structurally stable.*

Proof. Take $\varepsilon > 0$ and find for the hyperbolic set $I = M$ a neighborhood $W(\varepsilon)$ as in the proof of Theorem 12.8. For $\tilde{f} \in W(\varepsilon)$ find a homeomorphism h conjugating f on M and $\tilde{f}$ on $h(M)$. To complete the proof let us show that $h(M) = M$. It follows from Lemma 12.8 that for $\tilde{p} \in M$ there exists $p \in M$ such that

$$d(\tilde{f}^k(\tilde{p}), f^k(p)) < \varepsilon, \quad k \in \mathbf{Z},$$

then $\tilde{p} = h(p)$. Hence, $h(M) = M$. □

We are going now to prove Theorem 12.5.

Lemma 12.9. *Suppose that f satisfies Axiom A and has no 1-cycles. Then for any basic set Ω_i there exists a neighborhood U_i such that*

$$f^k(x) \in U_i, \quad k \in \mathbf{Z}, \tag{12.51}$$

implies $x \in \Omega_i$.

Remark. The property of Ω_i we are going to establish is sometimes called the local maximality of Ω_i.

Proof. The basic sets $\Omega_1, \ldots, \Omega_m$ are disjoint compacts so that there exist neighborhoods $U_i, \ldots, U_m$ such that

$$\overline{U}_i \cap \overline{U}_j = \emptyset, \quad i \neq j.$$

Let us show that the neighborhood U_i has the desired property. Take $x \in M$ such that (12.51) holds. It follows from Theorem 12.3 that there exists basic sets $\Omega_{i_1}, \Omega_{i_2}$ such that

$$x \in W^u(\Omega_{i_1}) \cap W^s(\Omega_{i_2})$$

i.e.,

$$d(f^k(x), \Omega_{i_1}) \underset{k\to-\infty}{\longrightarrow} 0, \quad d(f^k(x), \Omega_{i_2}) \underset{k\to+\infty}{\longrightarrow} 0.$$

We obtain from the choice of $U_1, \dots, U_m$ and from (12.51) that $\Omega_{i_1} = \Omega_{i_2} = \Omega_i$. As f has no 1-cycles, we see that $x \in \Omega_i$. □

Let us prove Theorem 12.4. As it was said above we prove this result supposing the absence of 1-cycles. Consider $x \in M$. Let us show that there exists $p \in \Omega$ such that $x \in W^s(p)$.

It follows from Theorem 12.3 that there exists a basic set Ω_i such that $x \in W^s(\Omega_i)$, i.e.,

$$d(f^k(x), \Omega_i) \underset{k\to+\infty}{\longrightarrow} 0.$$

Consider a neighborhood U_i of Ω_i having the property described in Lemma 12.9. Suppose that $\Omega_i \in H(C, \lambda)$. Find $\delta_0 > 0$ such that $\lambda_1 = \lambda + \delta_0 < 1$. Let $C_1 = C + \delta_0$. Find a neighborhood U_{δ_0} corresponding to δ_0 (see Theorem 12.7, we take f as $\tilde{f}$ in this theorem), and suppose that $U_{\delta_0} \subset U_0$.

Take $\varepsilon_1 > 0$ and $m > 0$ such that

$$f^k(x) \in U_{\delta_0}, \quad d(f^k(x), \Omega_i) < \varepsilon_1$$

for $k \geq m$. Let $\tilde{p} = f^m(x)$. Find a point $q \in \Omega_i$ such that $d(q, \tilde{p}) < \varepsilon_1$. It follows from the proof of Theorem 12.7 that the subspaces $\widetilde{L}^+(p), \widetilde{L}^-(p)$ are close to $L^+(q), L^-(q)$, respectively if ε_1 is small. Therefore if ε_1 is small enough then there exists a point p_1 such that

$$p_1 \in \widetilde{W}^s_{0,5\varepsilon_1}(\tilde{p}) \cap W^u(q)$$

and

$$d(f^k(p_1), f^k(\tilde{p})) \leq C_1 \lambda_1^k d(p_1, \tilde{p}), \quad k \geq 0, \tag{12.52}$$

$$d(f^k(p_1), f^k(q)) \leq C_1 \lambda_1^{-k} d(p_1, q), \quad k \leq 0. \tag{12.53}$$

It follows from (12.52), (12.53) that if ε_1 is small enough then

$$f^k(p_1) \in U_i, \quad k \in \mathbf{Z}.$$

We obtain that $p_1 \in \Omega_i$. Taking into account (12.52) we see that $\tilde{p} \in W^s(p_1)$. It is evident that

$$x \in W^s(p), p = f^{-m}(p_1).$$

This completes the proof of Theorem 12.4. □

Consider an open set $U \subset M$. Let $\Omega(\tilde{f})$ be the nonwandering set of a diffeomorphism $\tilde{f}$. Denote the following set:

$$\Omega(\tilde{f}, U) = \{x \in \Omega(\tilde{f}) : \tilde{f}^k(x) \in U, \ k \in \mathbf{Z}\}.$$

Lemma 12.10. *Suppose that a diffeomorphism f satisfies Axiom A and has no 1-cycles. Consider a basic set Ω_i of f. Given $\varepsilon > 0$ there exists a neighborhood $U_i^0(\varepsilon)$ of Ω_i in M and a neighborhood $W_i^0(\varepsilon)$ of f in $\mathit{Diff}^1(M)$ such that for any $\tilde{f} \in W_i^0(\varepsilon)$ there is a map $h : \Omega_i \to M$ having the following properties:*

(1) *h is a homeomorphism of Ω_i onto $\Omega(\tilde{f}, U_i^0(\varepsilon))$ conjugating f on Ω_i and $\tilde{f}$ on $\Omega(\tilde{f}, U_i^0(\varepsilon))$;*

(2) *$d(x, h(x)) < \varepsilon$ for $x \in \Omega_i$.*

Proof. Consider a neighborhood U_i of Ω_1 having the property described in Lemma 12.9. Take $\varepsilon > 0$ and construct, as in the proof of Theorem 12.8, a neighborhood $U_i^0(\varepsilon)$ of Ω_i and a neighborhood $W_i^0(\varepsilon)$ of f such that:

(1) the ε-neighborhood of $\overline{U_i^0(\varepsilon)}$ is a subset of U_i;

2) for any $p \in \Omega_i$ there exists $\tilde{p}$ such that

$$d(f^k(p), \tilde{f}^k(\tilde{p})) < \varepsilon, \ \tilde{f}^k(\tilde{p}) \in U_i^0(\varepsilon), \ k \in \mathbf{Z};$$

(3) for any $\tilde{p}$ such that $\tilde{f}^k(\tilde{p}) \in U_i^0(\varepsilon), k \in \mathbf{Z}$, there exists a point p such that

$$d(f^k(p), \tilde{f}^k(\tilde{p})) < \varepsilon, \ k \in \mathbf{Z}.$$

Construct as in the proof of Theorem 12.8 a homeomorphism h such that h conjugates f on Ω_i and $\tilde{f}$ on $h(\Omega_i)$ and such that $d(x, h(x)) < \varepsilon$ for $x \in \Omega_i$. Let us show that

$$h(\Omega_i) = \Omega(\tilde{f}, U_i^0(\varepsilon)). \tag{12.54}$$

It follows from Axiom A that periodic points of f are dense in Ω_i. As h conjugates f and $\tilde{f}$, it is easy to see that periodic points of $\tilde{f}$ are dense in $h(\Omega_i)$. The nonwandering set of $\tilde{f}$ is closed and contains periodic points of $\tilde{f}$; therefore $h(\Omega_i) \subset \Omega(\tilde{f})$. Hence,

$$h(\Omega_i) \subset \Omega(\tilde{f}, U_i^0(\varepsilon)). \tag{12.55}$$

Consider now a point $\tilde{p} \in \Omega(\tilde{f}, U_i^0(\varepsilon))$. It follows from the definition that $\tilde{f}^k(\tilde{p}) \in U_i^0(\varepsilon)$, $k \in \mathbf{Z}$. We obtain from the choice of neighborhoods that there exists a point p such that

$$f^k(p) \in U_i,\ d(f^k(p), \tilde{f}^k(\tilde{p})) < \varepsilon,\ k \in \mathbf{Z}.$$

By our choice of U_i, $p \in \Omega_i$, hence $\tilde{p} = h(p) \in h(\Omega_i)$. We proved that

$$\Omega(\tilde{f}, U_i^0(\varepsilon)) \subset h(\Omega_i).$$

It follows from this inclusion and from (12.55) that (12.54) holds. □

Lemma 12.11. *Suppose that a diffeomorphism f satisfies Axiom A and the no-cycle condition. Given a neighborhood $\tilde{U}$ of Ω there exists a neighborhood $\widetilde{W}$ of f in $Diff^1(M)$ such that for $\tilde{f} \in \widetilde{W}$ we have*

$$\Omega(\tilde{f}) \subset \tilde{U}.$$

Proof. Take a neighborhood $\tilde{U}$ of Ω. Consider neighborhoods $U_1, \dots, U_m$ of the basic sets $\Omega_1, \dots, \Omega_m$ of f having the property described in Lemma 12.9. We suppose that

$$\overline{U}_i \subset \tilde{U};\ \overline{f(U_i) \cup U_i} \cap \overline{U}_j = \emptyset,\ i \neq j. \tag{12.56}$$

To get a contradiction suppose that there exists a sequence of diffeomorphisms f_k and a sequence of points $x_k \in M$ such that

$$\rho_1(f_k, f) \underset{k \to \infty}{\longrightarrow} 0,\ x_k \in \Omega(f_k),\ x_k \notin \tilde{U}.$$

Let x_0 be a limit point of x_k, we suppose that $x_k \to x_0$ as $k \to \infty$. Then $x_0 \notin \widetilde{U}$. Find sequences $\xi_k \in M$ and $\tau(k)$ such that

$$d(\xi_k, x_k) < \frac{1}{k},\ d(f_k^{\tau(k)}(\xi_k), x_k) < \frac{1}{k},\ \tau(k) > k.$$

Define the following set

$$\Xi = \{x = \lim_{k\to\infty} f_k^{\theta(k)}(\xi_k) : 0 \le \theta(k) \le \tau(k)\}.$$

Let us show that Ξ is invariant with respect to f. Note for the beginning that

$$f^m(x_0) \in \Xi,\ \ m \in \mathbf{Z}.$$

Indeed, if $m \ge 0$ then

$$f^m(x_0) = \lim_{k\to\infty} f_k^m(\xi_k),$$

if $m < 0$ then

$$f^m(x_0) = \lim_{k\to\infty} f_k^m(f_k^{\tau(k)}(\xi_k)).$$

We take into account here that for $m \ge 0$ and for large k, we have $0 \le m \le \tau(k)$, and for $m < 0$ and for large k, we have $0 \le \tau(k)+m < \tau(k)$.

Consider $\xi \in \Xi$, $\xi = \lim f_k^{\theta(k)}(\xi_k)$. If one of the sequences $\theta(k)$, $\tau(k)-\theta(k)$ contains a sequence bounded from above then we can assume that $\theta(k) \to \theta$ as $k \to \infty$, then

$$\xi = \lim_{k\to\infty} f_k^{\theta(k)}(\xi_k) = f^\theta(x_0),$$

and we obtain that the trajectory of ξ belongs to Ξ. If $\theta(k) \to \infty, \tau(k)-\theta(k) \to \infty$ as $k \to \infty$, then for any $m \in \mathbf{Z}$ and for large k we have

$$0 \le \theta(k) + m \le \tau(k)$$

and hence

$$f^m(\xi) = \lim_{k\to\infty} f_k^{\theta(k)+m}(\xi_k) \in \Xi.$$

Evidently the set Ξ is closed. Consider the basic sets $\Omega^{(1)}, \dots, \Omega^{(s)}$ of f such that $\Omega^{(i)} \cap \Xi \ne \emptyset$ for $i = 1, \dots, s$.

It follows from Theorem 12.3 that there exist basic sets Ω', Ω'' such that $x_0 \in W^u(\Omega') \cap W^s(\Omega'')$. As the set Ξ is closed and invariant, we obtain that the sets Ω', Ω'' belong to $\{\Omega^{(1)}, \dots, \Omega^{(s)}\}$. Suppose for definiteness that $x_0 \in W^s(\Omega^{(1)})$. If

$$x_0 \in W^u(\Omega^{(1)}) \tag{12.57}$$

we obtain a contradiction between the choice of $x_0 \in \Omega$ and the absence of 1-cycles.

Consider the case in which (12.57) is not satisfied. Let $x_0 \in W^u(\Omega^{(2)})$. The set $\Omega^{(2)}$ is compact; therefore there exists $d_1 > 0$ such that

$$d(f^k(x_0), \Omega^{(2)}) \geq d_1$$

for $k \geq 0$. Passing if necessary to a subsequence of ξ_k we assume that

$$d(f_k^m(\xi_k), f^m(x_0)) < \frac{1}{k}, 0 \leq m \leq k.$$

Then $f_k^k(\xi_k) \underset{k\to\infty}{\longrightarrow} \Omega^{(1)}$ and for large k we have

$$d(f_k^m(\xi_k), \Omega^{(2)}) \geq \frac{d_1}{2}, \quad 0 \leq m \leq k. \tag{12.58}$$

Denote $\eta_k = f_k^k(\xi_k)$. We obtain from (12.58) and from the definition of Ξ that there exists $\alpha(k) > 0$ such that

$$d(f_k^{\alpha(k)}(\eta_k), \Omega^{(2)}) \underset{k\to\infty}{\longrightarrow} 0.$$

It is easy to see that $\alpha(k) \to \infty$ as $k \to \infty$. Consider neighborhoods $U^{(1)}, \dots, U^{(s)}$ of $\Omega^{(1)}, \dots, \Omega^{(s)}$ such that analogues of (12.56) are valid. For large k we have

$$\overline{f_k(U^{(i)}) \cup U^{(i)}} \cap \overline{U^{(j)}} = \emptyset, \quad i \neq j.$$

Consider the set of sequences of pairs $(\theta^-(k), \theta^+(k))$ such that

(1) $f_k^m(\eta_k) \in U^{(1)}, \theta^-(k) \leq m \leq \theta^+(k);$ (12.59)

(2) $\theta^+(k) - \theta^-(k) \underset{k\to\infty}{\longrightarrow} +\infty.$ (12.60)

Evidently this set is nonempty. For any k denote by $\theta^*(k)$ the largest $\theta^+(k)$ in pairs $(\theta^-(k), \theta^+(k))$.

Then we have

$$f_k^{\theta^*(k)}(\eta_k) \in U^{(1)}, \ \tilde{\eta}_k = f_k^{\theta^*(k)+1}(\eta_k) \notin U^{(1)}.$$

Let z_1 be a limit point of the sequence $\tilde{\eta}_k$. It is easy to see that

$$z_1 \in \Xi, \; z_1 \in \overline{f(U_1)} \backslash U_1. \tag{12.61}$$

We obtain from (12.59) and (12.60) that for any $m < 0$

$$f_k^{\theta^*(k)+1+m}(\eta_k) \in U^{(1)}.$$

Hence,

$$f^m(z_1) \in U^{(1)}, \; m < 0,$$

so that by Theorem 12.3 we have $z_1 \in W^u(\Omega^{(1)})$. There exists a basic set $\Omega^{(\sigma)} \in \{\Omega^{(1)}, \ldots, \Omega^{(s)}\}$ such that $z_1 \in W^s(\Omega^{(\sigma)})$. If $\sigma = 1$ we get a contradiction with the absence of 1-cycles (note that it follows from (12.61) that $z_1 \notin \Omega$); if $\sigma = 2$ we obtain a 2-cycle

$$\Omega^{(2)} \to \Omega^{(1)} \to \Omega^{(2)}.$$

If $\sigma \neq 1, 2$ we repeat the construction finding a point

$$z_2 \in W^u(\Omega^{(\sigma)})$$

being a limit point of the sequence $f^m(\eta_k), 0 \leq m \leq \alpha(k)$, and so on. It follows from the finiteness of the set $\{\Omega^{(1)}, \ldots, \Omega^{(s)}\}$ that by repeating the construction we obtain a cycle and hence get a contradiction. □

To complete the proof of Theorem 12.5 let us fix $\varepsilon > 0$. Find corresponding neighborhoods $U_i^0(\varepsilon)$ of $\Omega_1, \ldots, \Omega_m$, and neighborhoods $W_i^0(\varepsilon)$ of f (see Lemma 12.10). The set

$$\widetilde{U} = U_1^0(\varepsilon) \cup \cdots \cup U_m^0(\varepsilon)$$

is a neighorhood of Ω. There exists a corresponding neighborhood $\widetilde{W}$ of f constructed in Lemma 12.11. Consider the neighborhood

$$W = \widetilde{W} \cap W_1^0(\varepsilon) \cap \ldots \cap W_m^0(\varepsilon)$$

of f. For $\tilde{f} \in W$ let

$$\widetilde{\Omega} = \bigcup_{i=1}^{m} \Omega(\tilde{f}, U_i^0(\varepsilon)).$$

It follows from Lemma 12.10 that there exists a homeomorphism h conjugating f on Ω and $\tilde{f}$ on $\tilde{\Omega}$ and such that $d(x, h(x)) < \varepsilon$ for $x \in \Omega$. We obtain from Lemma 12.10 that $\Omega(\tilde{f}) = \tilde{\Omega}$. This completes the proof of Theorem 12.5. □

We give now an example, due to Thom, of an Anosov diffeomorphism. Consider the torus T^2 identifying on $\mathbf{R}^2$ points (x_1, y_1) and (x_2, y_2) if $x_1 - x_2 \in \mathbf{Z}$, $y_1 - y_2 \in \mathbf{Z}$.

Consider the matrix

$$A = \begin{pmatrix} 2 & 1 \\ 1 & 1 \end{pmatrix}.$$

As the entries of A are integers and det $A = 1$, the entries of A^{-1} are also integers. Hence the linear map $f_0(x) = Ax$ of the plane generates a diffeomorphism $f : T^2 \to T^2$. The eigenvalues of A are

$$\lambda_1 = \frac{3 + \sqrt{5}}{2}, \ \lambda_2 = \frac{3 - \sqrt{5}}{2}.$$

Evidently, $0 < \lambda_2 < 1 < \lambda_1$. At any point $p \in T^2$ identify in the natural way T_pT^2 with $\mathbf{R}^2$ and consider two linear subspaces of T_pT^2 : $L^+(p)$– the subspace spanned by the eigenvector v^+ corresponding to λ_2, and $L^-(p)$–the subspace spanned by the eigenvector v^- corresponding to λ_1. It follows from the equalities

$$|Df^k(p)v| = \lambda_2^k |v|, \ \ v \in L^+(p),$$
$$|Df^k(p)v| = \lambda_1^k |v|, \ \ v \in L^-(p),$$

that T^2 is a hyperbolic set of f, so f is an Anosov diffeomorphism. We obtain from Theorem 12.9 that diffeomorphism f is structurally stable. We leave it to the reader to show that periodic points of f are dense in T^2, and hence, $\Omega(f) = T^2$.

4. Let us formulate now the analogues of given definitions and results in the case of a system of differential equations.

Consider system (1.1) from $X_+^1(G)$, $G \subset \mathbf{R}^n$ (see Section 1, Chapter 3). We say that a set $I \subset \mathbf{R}^n$ is hyperbolic for (1.1) if:

(1) I is a compact invariant set for the flow φ generated by (1.1);

(2) there exists $C, \lambda > 0$ having the following property: for any $p \in I$ there exist two linear subspaces $L^+(p), L^-(p)$ of $\mathbf{R}^n$ such that

(a)

$$L^+(p) + L^-(p) + \{F(p)\} = \mathbf{R}^n.$$

Here $\{F(p)\}$ is the subspace spanned by the vector $F(p)$;

(b) if $\Phi(t,p)$ is the fundamental matrix of the system

$$\dot{y} = \frac{\partial F}{\partial x}(\varphi(t,p))y$$

such that $\Phi(0,p) = E$, then for any τ

$$\Phi(\tau,p)L^+(p) = L^+(\varphi(\tau,p)),$$
$$\Phi(\tau,p)L^-(p) = L^-(\varphi(\tau,p));$$

(c) for $v \in L^+(p)$, $t \geq 0$

$$|\Phi(t,p)v| \leq Ce^{-\lambda t}|v|;$$

(d) for $v \in L^-(p)$, $t \leq 0$

$$|\Phi(t,p)v| \leq Ce^{\lambda t}|v|.$$

The following condition is the analogue of Axiom A.

AXIOM A$'$.

1. The nonwandering set Ω of (1.1) is hyperbolic.
2. $\Omega = Q_1 \cup Q_2$; here Q_1 and Q_2 are disjoint compact invariant sets such that Q_1 is the union of a finite set of rest points, Q_2 contains no rest points, and closed orbits are dense in Q_2.

If (1.1) satisfies Axiom A$'$ then the following analogue of Theorem 12.2 is valid: there exists a unique decomposition

$$\Omega = \Omega_1 \cup \cdots \cup \Omega_m,$$

where Ω_i are compact disjoint invariant sets, and each Ω_i contains a dense trajectory.

The sets Ω_i are called basic sets. For two different basic sets Ω_i, Ω_j we write $\Omega_i \to \Omega_j$ if there exists a point x such that

$$\varphi(t,x) \underset{t\to-\infty}{\longrightarrow} \Omega_i, \ \varphi(t,x) \underset{t\to+\infty}{\longrightarrow} \Omega_j.$$

The no-cycle condition for system (1.1) repeats the no-cycle condition for the diffeomorphism.

Let us formulate the following analogue of Theorem 12.5.

Theorem 12.10. *If system* (1.1) *satisfies Axiom A′ and the no-cycle condition, then it is* Ω*-stable.*

Hyperbolic trajectories $\varphi(t,x)$ for $x \in \Omega$ have stable and unstable manifolds $W^s(\varphi(t,x))$, $W^u(\varphi(t,x))$.

We say that system (1.1) satisfies the geometric strong transversality condition if for any $x, y \in \Omega$, the manifolds $W^u(\varphi(t,x))$, $W^s(\varphi(t,x))$ are transversal.

The following statement is the analogue of Theorem 12.6.

Theorem 12.11 *If system* (1.1) *satisfies Axiom A′ and the geometric strong transversality condition then it is structurally stable.*

Similar statements are true for autonomous systems on a smooth closed manifold M.

Chapter 13

The Analytic Strong Transversality Condition

1. Let $f : M \to M$ be a diffeomorphism of a smooth closed n-dimensional manifold M. We introduced in Chapter 12 the geometric strong transversality condition for a diffeomorphism satisfying Axiom A: for any $p, q \in \Omega$ the manifolds $W^s(p)$, $W^u(q)$ are transversal.

Let us formulate now another strong transversality condition. For point $x \in M$ define two linear subspaces of T_xM:

$$B^+(x) = \{v \in T_xM : |Df^k(x)v| \underset{k \to +\infty}{\longrightarrow} 0\},$$
$$B^-(x) = \{v \in T_xM : |Df^k(x)v| \underset{k \to -\infty}{\longrightarrow} 0\}.$$

We say that f satisfies the analytic strong transversality condition if for any $x \in M$ we have

$$B^+(x) + B^-(x) = T_xM. \tag{13.1}$$

Theorem 13.1. *Suppose that a diffeomorphism f satisfies Axiom A. Then the geometric strong transversality condition is equivalent to the analytic strong transversality condition.*

Proof. Consider a point $x \in M$. It follows from Theorem 12.3 and from 12.4 that there exist basic sets Ω_i, Ω_j and points $q \in \Omega_j$, $p \in \Omega_i$ such that

$$x \in W^u(q) \cap W^s(p).$$

We claim that

$$B^+(x) = T_xW^s(p), \; B^-(x) = T_xW^u(q). \tag{13.2}$$

It follows from (13.2) that (13.1) is equivalent to the transversality of $W^s(p)$, $W^u(q)$ at x. As x is arbitrary, that will prove the theorem.

Let us establish the first equality in (13.2). The set Ω_i is hyperbolic. Suppose that $\Omega_i \in H(C,\lambda)$. Find ε_0 for the set Ω_i having the properties described in Theorem 12.1. Take $\varepsilon_1, \delta > 0$ such that

$$\lambda + \delta < 1,\ \varepsilon_1 + \delta < 0,5\varepsilon_0,\ 2(C+2)\varepsilon_1 < \varepsilon_0. \tag{13.3}$$

Find a neighborhood U_δ of Ω_i corresponding to δ (see Theorem 12.7 in which we take f as $\tilde{f}$). As $x \in W^s(p)$ there exists $k_+ > 0$ such that

$$f^k(x) \in U_\delta,\ d(f^k(x), f^k(p)) < \varepsilon_1 \tag{13.4}$$

for $k \geq k_+$. Define $x_0 = f^{k_+}(x)$, $r = f^{k_+}(p)$. It follows from (13.4) that

$$f^k(x_0) \in U_\delta,\ d(f^k(x_0), f^k(r)) < \varepsilon_1 \tag{13.5}$$

for $k \geq 0$. Let, as above, $D_{\varepsilon_1}(x_0)$ be the ball of radius ε_1 centered at x_0. We obtain from the second inequality in (13.3) that there exists a smooth disc $\widetilde{W}^s_{0,5\varepsilon_0}(x_0)$ having the properties described in Theorem 12.7.

Define the set

$$R(x_0) = D_{\varepsilon_1}(x_0) \cap \widetilde{W}^s_{0,5\varepsilon_0}(x_0).$$

If $x_1 \in R(x_0)$ we obtain from (12.8), (13.3) that for $k \geq 0$

$$d(f^k(x_1), f^k(x_0) \leq (C+\delta)(\lambda+\delta)^k d(x_1, x_0) < (C+\delta)\varepsilon_1.$$

Taking into account (13.5) and the inequality $\delta < 1$ we see that

$$d(f^k(x_1), f^k(r)) < (C+2)\varepsilon_1 < 0,5\varepsilon_0.$$

Then $x_1 \in W^s_{0,5\varepsilon_0}(r) \subset W^s(r)$, and hence

$$R(x_0) \subset W^s(r). \tag{13.6}$$

It follows from (13.6) that

$$\widetilde{L}^+(x_0) = T_{x_0}\widetilde{W}^s_{0,5\varepsilon_0}(x_0) \subset T_{x_0}W^s(r) \tag{13.7}$$

We show similarly that

$$D_{\varepsilon_1}(r) \cap W^s_{0,5\varepsilon_0}(r) \subset \widetilde{W}^s_{0,5\varepsilon_0}(x_0)$$

and

$$T_{x_0}W^s(r) \subset \widetilde{L}^+(x_0). \tag{13.8}$$

We obtain from (13.7) and (13.8) that

$$\widetilde{L}^+(x_0) = T_{x_0}W^s(r). \tag{13.9}$$

It follows from (12.6) that for $v \in \widetilde{L}^+(x_0)$ we have

$$|Df^k(x_0)v| \underset{k\to+\infty}{\longrightarrow} 0$$

so that

$$\widetilde{L}^+(x_0) \subset B^+(x_0). \tag{13.10}$$

Let us show that

$$B^+(x_0) \subset \widetilde{L}^+(x_0). \tag{13.11}$$

In order to get a contradiction suppose that there exists a vector $v \in B^+(x_0)\backslash \widetilde{L}^+(x_0)$. Then

$$v = v_1 + v_2,\ v_1 \in \widetilde{L}^+(x_0),\ v_2 \in \widetilde{L}^-(x_0)$$

and $v_2 \neq 0$. Denote $C_1 = C + \delta$, $\lambda_1 = \lambda + \delta$. As

$$|Df^k(x_0)v| \underset{k\to+\infty}{\longrightarrow} 0$$

there exists $a > 0$ such that

$$|Df^k(x_0)v| \leq a$$

for $k \geq 0$. From

$$|Df^k(x_0)v_1| \leq C_1\lambda_1^k|v_1| \underset{k\to+\infty}{\longrightarrow} 0$$

we obtain that there exists $b > 0$ such that

$$|Df^k(x_0)v_2| \leq b \tag{13.12}$$

for $k \geq 0$. We can write the inequality (12.3) as

$$|v_2| \leq C_1\lambda_1^k|Df^k(x_0)v_2| \leq C_1\lambda_1^k b \tag{13.13}$$

for $k \geq 0$. Passing to the limit in (13.13) for $k \to +\infty$ we see that $v_2 = 0$. The obtained contradiction proves (13.11). It follows from (13.9), (13.10), and (13.11) that

$$T_{x_0}W^s(r) = B^+(x_0).$$

Applying $Df^{-k_+}(x_0)$ to the last equality we obtain the first equality in (13.12). The second equality in (13.12) is established similarly. □

R. Mañé showed that the analytic strong transversality condition is a "stronger" form of the transversality condition than the geometric one [12]. He proved the following result.

Theorem 13.2 *If a diffeomorphism f satisfies the analytic strong transversality condition then f satisfies Axiom A.*

By Theorems 12.6, 13.1 and 13.2 the analytic strong transversality condition implies structural stabilty. Axiom A consists of two requirements on f: the nonwandering set Ω of f is hyperbolic and the periodic points of f are dense in Ω. We are going to show in the remaining sections of this chapter that the analytic strong transversality condition implies the hyperbolicity of Ω.

2. Let TM be the tangent bundle of the manifold (see Section 4 of Chapter 1). Define the following map $\pi : TM \to TM$ related to the diffeomorphism f: for $(x, v) \in TM$ $\pi(x, v) = (f(x), Df(x)v)$.

Let us call a subbundle of TM a set Y of pairs (x, Y_x) where every Y_x is a linear subspace of T_xM. We say that a subbundle Y is invariant with respect to π if for any $x \in M$

$$Df(x)Y_x = Y_{f(x)}.$$

We suppose now that f satisfies the analytic strong transversality condition. Consider two subbundles B^+, B^- connected with the decomposition (13.1) in the following natural way:

$$B_x^+ = B^+(x), B_x^- = B^-(x).$$

As

$$|Df^k(x)v| \underset{k\to+\infty}{\longrightarrow} 0$$

if and only if

$$|Df^k(f(x))Df(x)v| \underset{k\to+\infty}{\longrightarrow} 0,$$

we see that the subbundle B^+ is invariant with respect to π. Similarly B^- is also invariant with respect to π.

The main construction we are going to use is the map π^* dual to π. To define π^*, denote by $<,>$ the scalar product in T_xM.

Define the map $D^*f(x) : T_{f(x)}M \to T_xM$ so: for any $v \in T_xM$, $\xi \in T_{f(x)}M$

$$< \xi, Df(x)v >=< D^*f(x)\xi, v >$$

(i.e. $D^*f(x)$ is adjoint to $Df(x)$). Define now π^*: for $(f(x), \xi)$ where $\xi \in T_{f(x)}M$ let

$$\pi^*(f(x), \xi) = (x, D^*f(x)\xi).$$

Consider the projection $p : TM \to M$ defined in Chapter 1, $p(x, v) = x$. It is easy to see that

$$p(\pi(x, v)) = f(x),\ p(\pi^*(x, v)) = f^{-1}(x).$$

We obtain immediately from the definition of π^* that the following statement is true.

Lemma 13.1. $(\pi^*)^* = \pi$.

Consider a subbundle Y of TM. Define the subbundle $Y^\perp$ orthogonal to Y in the following way: for $x \in M$

$$Y_x^\perp = \{\xi :< \xi, v >= 0 \text{ for any } v \in Y_x\}.$$

Lemma 13.2. *If a subbundle Y is invariant with respect to π then $Y^\perp$ is invariant with respect to π^*.*

Proof. Consider vectors $\xi \in Y_{f(x)}^\perp$ and $D^*f(x)\xi \in T_xM$. For any $v \in Y_x$ we have

$$< v, D^*f(x)\xi >=< \xi, Df(x)v >= 0$$

as $Df(x)v \in Y_{f(x)}$. So $D^*f(x)\xi \in Y_x^\perp$. □

We say that two subbundles Y^1, Y^2 are complementary if for any $x \in M$ we have

$$T_xM = Y_x^1 \oplus Y_x^2, \tag{13.14}$$

where $\oplus$ denotes the direct sum.

Lemma 13.3. *Let Y^1, Y^2 be complementary subbundles invariant with respect to π. Then $(Y^1)^\perp, (Y^2)^\perp$ are complementary subbundles invariant with respect to π^*.*

Proof. By Lemma 13.2, $(Y^1)^\perp$ and $(Y^2)^\perp$ are π^*-invariant. Let $\dim Y_x^1 = k$. It follows from (13.14) that $\dim Y_x^2 = n - k$. Then it is evident that

$$\dim (Y_x^1)^\perp = n - k, \ \dim (Y_x^2)^\perp = k. \tag{13.15}$$

Consider a vector $\xi \in (Y_x^1)^\perp \cap (Y_x^2)^\perp$. We obtain from (13.14) that we can write an arbitrary vector $v \in T_xM$ as

$$v = v_1 + v_2, \ v_1 \in Y_x^1, \ v_2 \in Y_x^2.$$

Then $< \xi, v >=< \xi, v_1 > + < \xi, v_2 >= 0$. As v is arbitrary we see that $\xi = 0$. The equality

$$(Y_x^1)^\perp \cap (Y_x^2)^\perp = \{0\}$$

and (13.15) imply that $(Y_x^1)^\perp$, $(Y_x^2)^\perp$ are complementary. □

Consider an invariant set $M_0 \subset M$ for f. If M_0 is a hyperbolic set of f then on M_0 there are two complementary subbundles L^+, L^- that are invariant with respect to π and such that for $p \in M_0, L_p^+ = L^+(p), L_p^- = L^-(p)$, inequalities (12.3) and (12.4) hold.

Let us say in this case that the set M_0 is hyperbolic with respect to π with subbundles L^+, L^- and with constants C, λ.

Lemma 13.4. *If M_0 is hyperbolic with respect to π with subbundles L^+, L^- and with constants C, λ, then M_0 is hyperbolic with respect to π^* with subbundles $(L^+)^\perp, (L^-)^\perp$ with the same constants.*

Proof. It is well-known that for linear operators $A, B (AB)^* = B^*A^*$, hence for any $x \in M$

$$(Df(f(x))Df(x))^* = D^*f(x))D^*f(f(x)).$$

Using induction it is easy to show that for any $k \in \mathbf{Z}$, $v \in T_x$, M, $\xi \in T_{f^k(x)}M$ we have

$$< v, D^* f^k \xi > = < \xi, Df^k v > . \tag{13.16}$$

Here,

$$D^* f^k = D^* f(x) D^* f(f(x)) \cdots D^* f(f^{k-1}(x)),$$
$$Df^k = Df(f^{k-1}(x)) Df(f^{k-2}(x)) \cdots Df(x).$$

Consider subbundles $(L^+)^\perp, (L^-)^\perp$. By Lemma 13.3 these subbundles are complementary and invariant with respect to π^*.

Take $k > 0$ and $\xi \in (L^-_{f^k(x)})^\perp$. It is easy to see that $D^* f^k \xi \in T_x M$. We can write

$$|\eta| = \max_{|v|=1} < \eta, v >$$

for $\eta, v \in T_x M$. We have

$$|D^* f^k \xi| = \max_{|v|=1} < v, D^* f^k \xi > .$$

Write $v = v_1 + v_2$, where $v_1 \in L^+_x$, $v_2 \in L^-_x$. As $(L^-)^\perp$ is π^*-invariant, we have

$$D^* f^k \xi \in (L^-_x)^\perp,$$

and hence $< D^* f^k \xi, v_2 > = 0$. Therefore,

$$|D^* f^k \xi| = \max_{|v|=1} < v_1, D^* f^k \xi > = \max_{|v_1|=1} < \xi, Df^k v_1 > \le C\lambda^k |\xi|.$$

We take into account here (12.3) and the inequality $< \xi, \eta > \le |\xi|\,|\eta|$. Similar arguments show that for $k \le 0$, $\xi \in (L^+_{f^k(x)})^\perp$ we have

$$|D^* f^k \xi| \le C\lambda^{-k} |\xi|. \qquad \square$$

Suppose now that f satisfies the analytic strong transversality condition. For $(x, v) \in TM, k \in \mathbf{Z}$, denote $(x_k, v_k) = (\pi^*)^k (x, v)$.

Lemma 13.5. *For $(x, v) \in TM$ the inequality*

$$\sup_{k \in \mathbf{Z}} |v_k| < +\infty \tag{13.17}$$

implies $v = 0$.

Proof. Take $x \in M; \xi, u \in T_xM; k \in \mathbf{Z}$. Using evident equalities $x = f^{-k}(f^k(x)), u = Df^{-k}(Df^k u)$ we obtain that

$$<\xi, u> = <\xi, Df^{-k}(Df^k u)> = <D^* f^{-k}\xi, Df^k u>.$$

Suppose that (13.17) holds for $(x, v) \in TM$. It follows from (13.1) that we can write $\xi = \xi_1 + \xi_2$ so that

$$|Df^l \xi_1| \underset{l \to +\infty}{\longrightarrow} 0, \quad |Df^m \xi_2| \underset{m \to -\infty}{\longrightarrow} 0.$$

Then

$$\begin{aligned} <v, \xi> &= <v, \xi_1 + \xi_2> \\ &= <v, Df^{-l}(Df^l \xi_1)> + <v, Df^{-m}(Df^m \xi_2)> \\ &= <D^* f^{-l} v, Df^l \xi_1> + <D^* f^{-m} v, Df^m \xi_2>. \end{aligned}$$

It follows from (13.17) that $|D^* f^{-l} v|, |D^* f^{-m} v|$ are bounded. Hence, the first term tends to zero as $l \to +\infty$, the second term tends to zero as $m \to -\infty$. We see that $<v, \xi> = 0$ for any ξ, therefore $v = 0$.

Let us say that π^* satisfies condition B if for π^* the statement of Lemma 13.5 is fulfilled. Following [33] we study in Section 3 maps satisfying condition B.

3. For convenience denote π^* by ρ and write

$$\rho(x, v) = (\varphi(x), \Phi(x)v),$$

i.e. $\varphi(x) = f^{-1}(x), \Phi(x) = D^* f(x)) : T_xM \to T_{\varphi(x)}M$. Denote also

$$\Phi(0, x) = E : T_xM \to T_xM,$$

for $k > 0$

$$\begin{aligned} \Phi(k, x) &= \Phi(\varphi^{k-1}(x))\Phi(\varphi^{k-2}(x)) \cdots \Phi(x), \\ \Phi(-k, x) &= \Phi^{-1}(\varphi^{-k+1}(x)) \cdots \Phi^{-1}(x). \end{aligned}$$

It follows from our previous considerations that the map ρ is continuous and satisfies condition B in the following form: if for $(x, v) \in TM$ we have

$$\sup_{k \in \mathbb{Z}} |\Phi(k, x)v| < +\infty$$

then $v = 0$.

We sometimes denote $\Phi(t, x)$, $t \in \mathbf{Z}$, instead of $\Phi(k, x)$. Define two subbundles of $TM : S = \{(x, S_x)\}, U = \{(x, U_x)\}$ as follows:

— $v \in T_x M$ belongs to S_x, if $|\Phi(k, x)v| \underset{k \to +\infty}{\longrightarrow} 0$,

— $v \in T_x M$ belongs to U_x, if $|\Phi(k, x)v| \underset{k \to -\infty}{\longrightarrow} 0$.

It is easy to see that the subbundles S, U are invariant with respect to ρ.

Lemma 13.6. *Suppose that a sequence $(x_m, v_m) \in TM$ has the following properties:*

(1) $(x_m, v_m) \to (\tilde{x}, \tilde{v})$ *as* $m \to \infty$;

(2) *there exists* $L > 0$ *and a sequence* $t_m \underset{m \to \infty}{\longrightarrow} +\infty$ *such that*

$$|\Phi(t, x_m)v_m| \leq L \tag{13.18}$$

for $0 \leq t \leq t_m$. *Then* $(\tilde{x}, \tilde{v}) \in S$.

Proof. Take arbitrary $k_0 \geq 0$. There exists m_0 such that $t_m > k_0$ for $m \geq m_0$. Using (13.18) we obtain that

$$|\Phi(k_0, x_m)v_m| \leq L. \tag{13.19}$$

It follows from the continuity of ρ that $\Phi(k_0, x)v$ is continuous with respect to x, v. Passing to the limit in (13.19) we see that

$$|\Phi(k_0, \tilde{x})\tilde{v}| \leq L.$$

Hence, for any $k \geq 0$

$$|\Phi(k, \tilde{x})\tilde{v}| \leq L. \tag{13.20}$$

Consider an ω-limit point (x_0, v_0) of the sequence $(\varphi^t(\tilde{x}), \Phi(t, \tilde{x})\tilde{v})$, i.e. the limit of a sequence

$$(\varphi^{\tau_m}(\tilde{x}), \Phi(\tau_m, \tilde{x})\tilde{v}) \tag{13.21}$$

for some sequence $\tau_m \to +\infty$.

Fix $k \in \mathbf{Z}$. As

$$\varphi^{\tau_m}(\tilde{x}) \underset{m \to \infty}{\longrightarrow} x_0, \; \Phi(\tau_m, \tilde{x})\tilde{v} \underset{m \to \infty}{\longrightarrow} v_0$$

we obtain that

$$\varphi^{k+\tau_m}(\tilde{x}) \underset{m\to\infty}{\longrightarrow} \varphi^k(x_0), \ \Phi(k+\tau_m, \tilde{x})\tilde{v} \underset{m\to\infty}{\longrightarrow} \Phi(k, x_0)v_0. \qquad (13.22)$$

For large m we have $k+\tau_m > 0$, so that from (13.20), (13.22) it follows that

$$|\Phi(k, x_0)v_0| \le L. \qquad (13.23)$$

As (13.23) is fulfilled for any k, we apply condition B and see that $v_0 = 0$. So, in any convergent sequence (13.21) such that $\tau_m \to +\infty$ we have

$$|\Phi(\tau_m, \tilde{x})\tilde{v}| \to 0$$

so, $(\tilde{x}, \tilde{v}) \in S$. □

Remark. Similar arguments show that if $t_m \to -\infty$ in the statement of Lemma 13.6, then $(\tilde{x}, \tilde{v}) \in U$. We study below the case of the subbundle S; for U the statements are analogous.

Define the following set:

$$A = \{(x, v) \in TM : |\Phi(k, x)v| \le 1 \ \text{for}\, k \ge 0\}.$$

We say that a set $C = \{(x, v) \in TM\}$ is bounded if

$$\sup_{(x,v)\in C} |v| < +\infty.$$

As M is compact, it is easy to see that any closed and bounded subset C of TM is compact.

Lemma 13.7. *The set A is a compact subset of S.*

Proof. It was shown in the proof of Lemma 13.6 that if (13.20) is fulfilled for any $k \ge 0$ then $(\tilde{x}, \tilde{v}) \in S$, and hence $A \subset S$. As $\Phi(0, x)v = v$, A is bounded. Consider a sequence $(x_m, v_m) \in A$ such that $(x_m, v_m) \underset{m\to\infty}{\longrightarrow} (x_0, v_0)$. For any $k \ge 0$

$$|\Phi(k, x_0)v_0| = \lim_{m\to\infty} |\Phi(k, x_m)v_m| \le 1,$$

and hence $(x_0, v_0) \in A$ so that A is closed. □

Lemma 13.8. *Given $\lambda > 0$ there exists $T > 0$ such that for any $(x, v) \in A$ we have*

$$|\Phi(t, x)v| < \lambda \tag{13.24}$$

for $t \geq T$.

Proof. To get a contradiction suppose that there exists $\lambda > 0$ and sequences $(x_m, v_m) \in A$, $t_m \to +\infty$ such that

$$|\Phi(t_m, x_m)v_m| \geq \lambda. \tag{13.25}$$

As A is compact, there exists a convergent subsequence of $(\varphi^{t_m}(x_m), \Phi(t_m, x_m)v_m) \in A$. Suppose for definiteness that

$$(\varphi^{t_m}(x_m), \Phi(t_m, x_m)v_m) \underset{m\to\infty}{\longrightarrow} (\tilde{x}, \tilde{v}).$$

It follows from (13.25) that $|\tilde{v}| \geq \lambda$. Take $k \in \mathbf{Z}$. For large m we have $t_m + k > 0$. As

$$(\varphi^{t_m+k}(x_m), \Phi(t_m + k)x_m)v_m) \underset{m\to\infty}{\longrightarrow} (\varphi^k(\tilde{x}), \Phi(k, \tilde{x})\tilde{v})$$

and $|\Phi(t_m + k, x_m)v_m| \leq 1$ we see that $|\Phi(k, \tilde{x})\tilde{v}| \leq 1$. As k is arbitrary, we obtain from condition B that $\tilde{v} = 0$. The contradiction completes the proof. □

Lemma 13.9. *There exists $\lambda_0 > 0$ such that for $(x, v) \in S$ the inequality $|v| \leq \lambda_0$ implies $(x, v) \in A$.*

Proof. To get a contradiction suppose that there exists a sequence $(x_m, v_m) \in S$ such that

$$|v_m| \underset{m\to\infty}{\longrightarrow} 0, \ (x_m, v_m) \notin A.$$

Let

$$\mu_m = \max_{k\geq 0} |\Phi(k, x_m)v_m| > 1.$$

For any m find t_m such that

$$|\Phi(t_m, x_m)v_m| = \mu_m.$$

As

$$|\Phi(k, x_m)\frac{v_m}{\mu_m}| \leq 1$$

for $k \geq 0$, we obtain that $(x_m, \frac{v_m}{\mu_m}) \in A$. It follows from the continuity of ρ and from the equality $\Phi(t, x)0 = 0$ for any t, x, that for any $T > 0$

$$\max_{0 \leq k \leq T} |\Phi(k, x_m)\frac{v_m}{\mu_m}| \underset{m \to \infty}{\longrightarrow} 0.$$

We take into account here that $|v_m| \to 0$, $\mu_m > 1$. Hence, $t_m \to +\infty$ as $m \to \infty$. As

$$(x_m, \frac{v_m}{\mu_m}) \in A,\ t_m \to +\infty,\ |\Phi(t_m, x_m)\frac{v_m}{\mu_m}| = 1$$

we obtain a contradiction with Lemma 13.8. □

Lemma 13.10. *There exists $T > 0$ such that for any $(x, v) \in S$*

$$|\Phi(t, x)v| \leq \frac{|v|}{2} \tag{13.26}$$

for $t \geq T$.

Proof. By Lemma 13.8 there exists $T > 0$ such that for any $(x, \tilde{v}) \in A$

$$|\Phi(t, x)\tilde{v}| \leq \frac{\lambda_0}{2}$$

for $t \geq T$. Here λ_0 is given by Lemma 13.9. Consider arbitrary $(x, v) \in S$, and suppose that $v \neq 0$. Then

$$(x, \tilde{v}) \in A, \text{ where } \tilde{v} = \lambda_0 \frac{v}{|v|}.$$

For $t \geq T$ we have

$$|\Phi(t, x)\tilde{v}| = \frac{\lambda_0}{|v|}|\Phi(t, x)v| \leq \frac{\lambda_0}{2}.$$

This inequality is equivalent to (13.26). In the case $v = 0$ (13.26) is evident. □

Lemma 13.11 (1). *Subbundles S, U are closed;* (2) *there exists $C > 0$, $\alpha \in (0,1)$ such that*

$$|\Phi(t,x)v| \leq C\alpha^t|v| \tag{13.27}$$

for $(x,v) \in S$, $t \geq 0$;

$$|\Phi(t,x)v| \leq C\alpha^{-t}|v| \tag{13.28}$$

for $(x,v) \in U$, $t \leq 0$.

Proof. We consider only the case of subbundle S; the case of U is treated similarly.

Consider a sequence $(x_k, v_k) \in S$ such that $(x_k, v_k) \to (x,v)$ as $k \to \infty$. If $v = 0$ evidently $(x,v) \in S$. If $v \neq 0$ then $v_k \neq 0$ for large k. By Lemma 13.9 there exists $\lambda_0 > 0$ such that

$$(x_k, \lambda_0 \frac{v_k}{|v_k|}) \in A.$$

As A is closed (see Lemma 13.7) we obtain that

$$(x, \lambda_0 \frac{v}{|v|}) \in A.$$

By Lemma 13.7 $(x,v) \in S$. So the first statement is proved.

To prove the second statement consider $T > 0$ given by Lemma 13.10. It is easy to see that (13.26) implies the following inequalities:

$$|\Phi(2T,x)v| \leq (\frac{1}{2})^2|v|,$$

$$\dots$$

$$|\Phi(kT,x)v| \leq (\frac{1}{2})^k|v| \tag{13.29}$$

for any $(x,v) \in S$, $k > 0$. There exists $C_0 > 0$ such that

$$\max_{x \in M} (\|\Phi(0,x)\|, \|\Phi(1,x)\|, \dots, \|\Phi(T-1,x)\|) \leq C_0. \tag{13.30}$$

Here, as usual $\|\Phi(t,x)\| = \max_{|v|=1}|\Phi(t,x)v|$. Denote $C = 2C_0, \alpha = 2^{-1/T}$. We claim that (13.27) with these C, α is true.

We will now consider arbitrary $t \geq 0$. There exists $k \geq 0$ such that $t \in [kT, (k+1)T)$ i.e. $t = kT + \tau$ where $0 \leq \tau < T$. For $(x, v) \in S$ we conclude from (13.29) and (13.30) that

$$|\Phi(t,x)v| = |\Phi(\tau, \varphi^{kT}(x))\Phi(kT,x)v| \leq \frac{C_0}{2^k}|v|.$$

As $k+1 > \frac{t}{T}$, $-k < 1 - \frac{t}{T}$, $2^{-k} < 2\alpha^t$, we see that

$$|\Phi(t,x)v| \leq C\alpha^t|v|,$$

so that (13.27) is true. □

Remark. The inequalities (13.27) and (13.28) are of the same form as the inequalities (12.1)and (12.2) in the definition of a hyperbolic set. To prove that ρ is hyperbolic on a subset M_0 of M with subbundles S, U we will show now that

$$S_x + U_x = T_xM \tag{13.31}$$

for $x \in M_0$.

Lemma 13.12 *Consider sequences* $(x_m, v_m) \in TM$, $t_m \to +\infty$, *and suppose that there exists* $r > 0$ *such that*

$$|v_m| \leq r,\ |\Phi(t_m, x_m)v_m| \leq r \tag{13.32}$$

Then there exists $R > 0$ *such that*

$$|\Phi(t, x_m)v_m| \leq R$$

for $0 \leq t \leq t_m$. **Proof.** To get a contradiction suppose that there exists $(x_m, v_m) \in TM$, $t_m \to +\infty$, $r > 0$ such that (13.32) holds and the numbers

$$\beta_m = \max_{0 \leq t \leq t_m} |\Phi(t, x_m)v_m|$$

have the following property: $\beta_m \to \infty$ as $m \to \infty$. Define numbers $\tau_m \in [0, t_m]$ by: $\beta_m = |\Phi(\tau_m, x_m)v_m|$. It follows immediately from the continuity of ρ that

$$\tau_m \underset{m\to\infty}{\longrightarrow} +\infty,\ t_m - \tau_m \underset{m\to\infty}{\longrightarrow} +\infty. \tag{13.33}$$

Let

$$w_m = \Phi(\tau_m, x_m)\frac{v_m}{\beta_m}.$$

Evidently $|w_m| = 1$. Consider a limit point (x, w) of the sequence $(\varphi^{\tau_m}(x_m), w_m)$. Note that $|w| = 1$. By Lemma 13.6 we obtain from the inequality

$$|\Phi(t, \varphi^{\tau_m}(x_m)w_m| \leq 1$$

where $t \in [-\tau_m, t_m - \tau_m]$ and from (13.33) that $w \in U_x \cap S_x$. It follows from condition B that $w = 0$. With this contradiction we complete the proof. □

Remark. An analogous statement is true if $t_m \to -\infty$. Then $|\Phi(t, x_m)v_m| \leq R$ for $t \in [t_m, 0]$.

Theorem 13.3. *If x is a nonwandering point of the diffeomorphism φ then (13.31) holds.*

Proof. It follows from the definition of a nonwandering point that there exist sequences $x_m \in M$, $t_m \in \mathbb{R}$ such that

$$x_m \to x,\ \varphi^{t_m}(x_m) \to x,\ |t_m| \to \infty$$

as $m \to \infty$. We can choose these sequences so that $t_m \to -\infty$.

Consider the linear subspace U_x of T_xM. Let Q be the orthogonal complement of U_x. Denote $s = \dim Q$. Fix an orthogonal basis $v_1, \ldots, v_s$ of Q. There exists a neighborhood V of x in M and a diffeomorphism β mapping the set

$$\{(\xi, w) : \xi \in V,\ w \in T_\xi M\}$$

onto $V \times \mathbb{R}^n$ (see Chapter 1). Using the coordinates generated by β choose s orthonormal vectors $v_1^m, \ldots, v_s^m$ in $T_{x_m}M$ such that $v_j^m \to v_j$ as $m \to \infty$, $j = 1, \ldots, s$. Denote by Q_m the linear subspace of $T_{x_m}M$ spanned by $v_1^m, \ldots, v_s^m$. Define numbers

$$\mu_m = \min\{|\Phi(t_m, x_m)v| : v \in Q_m, |v| = 1\}.$$

We claim that

$$\mu_m \underset{m\to\infty}{\longrightarrow} \infty. \tag{13.34}$$

To get a contradiction suppose that (13.34) is not true. Then there exists $r > 0$ and sequences $v_m \in Q_m$, $|v_m| = 1$, $t_m \underset{m\to\infty}{\longrightarrow} -\infty$ such that

$$|\Phi(t_m, x_m)v_m| \le r.$$

By the Remark to Lemma 13.12 there exists $R > 0$ such that

$$|\Phi(t, x_m)v_m| \le R$$

for $t \in [t_m, 0)$. By Lemma 13.6 we conclude that any limit point $(x, \tilde{v})$ of the sequence (x_m, v_m) belongs to U; hence $\tilde{v} \in U_x$. As $v_m \in Q_m$ the vector $\tilde{v}$ is orthogonal to U_x so that we obtain a contradiction with the construction. Therefore, (13.34) is established.

Define now the following linear space:

$$K_m = \Phi(t_m, x_m)Q_m.$$

Evidently $K_m \subset T_{y_m}M$, where $y_m = \varphi^{t_m}(x_m)$, $\dim K_m = s$. Consider a vector $w \in K_m$, $|w| = 1$. Let $w = \Phi(t_m, x_m)v$, then we obtain from the definition of μ_m that

$$|w| \ge \mu_m |v|, \ |v| \le \frac{1}{\mu_m}. \tag{13.35}$$

By Lemma 13.12 it follows from (13.34) and (13.35) that for any sequence (y_m, w_m) where $w_m \in K_m$ and $|w_m| = 1$ there exists $R > 0$ such that

$$|\Phi(t, y_m)w_m| \le R$$

for $t \in [0, -t_m]$. It follows now from Lemma 13.6 that any limit point $(x, \tilde{w})$ of (x_m, w_m) belongs to S so that $\tilde{w} \in S_x$.

Choose an orthognormal basic $w_1^m, \dots, w_s^m$ of K_m. For definiteness we suppose that

$$w_1^m \to \tilde{w}_1, \dots, w_s^m \to \tilde{w}_s$$

as $m \to \infty$. The vectors $\tilde{w}_1, \dots, \tilde{w}_s$ are orthonormal unit vectors in S_x. Hence

$$\dim S_x \ge s. \tag{13.36}$$

It follows from the definitions of Q, Q_m that

$$\dim U_x = n - s.$$

We obtain from (13.36) now that

$$\dim S_x + \dim U_x \geq n.$$

By condition B, $S_x \cap U_x = \{0\}$, hence

$$S_x + U_x = T_x M.$$

The nonwandering set of f coincides with the nonwandering set of $\varphi = f^{-1}$. We conclude now from Lemma 13.1 and from Lemma 13.4 applied to π^* that the analytic strong transversality condition for f implies the hyperbolicity of the nonwandering set of f.

Appendix

Proof of the Grobman-Hartman Theorem

1. Let us prove Theorem 4.3. We begin by proving the following statement concerning diffeomorphisms.

Theorem. *Let Φ be a diffeomorphism of class C^1 of a neighborhood U_0 of the origin of $\mathbf{R}^n$ onto a neighborhood of the origin such that $\Phi(0) = 0$. Suppose that $\mathbf{R}^n$ is decomposed: $\mathbf{R}^n = \mathbf{R}^{n_1} \times \mathbf{R}^{n_2}$ (with coordinates y in $\mathbf{R}^{n_1}$, z in $\mathbf{R}^{n_2}$) so that*

$$\Phi(x) = Lx + f(x) = (Ay + f_1(y, z), Bz + f_2(y, z)). \tag{A.1}$$

Here L, A, B, are constant matrices,

$$\|A\| < 1, \|B^{-1}\| < 1; \tag{A.2}$$

$$f = (f_1, f_2),\ f(0) = 0; \tag{A.3}$$

$$\frac{\partial f}{\partial x}(0) = 0. \tag{A.4}$$

Then the fixed point $x = 0$ of Φ is locally topologically conjugate to the fixed point $x = 0$ of the linear map $\Phi(x) = Lx$.

Proof. We are going to prove that there exists a neighborhood U of the origin of $\mathbf{R}^n$ and homeomorphisms h mapping U onto $h(U)$ and such that $h(0) = 0$ and

$$\Phi(h(x)) = h(\Psi(x)) \tag{A.5}$$

for $x \in U \cap h^{-1}(U)$. According to the decomposition $x = (y, z)$ the matrix L is block diagonal:

$$L = \begin{pmatrix} A & 0 \\ 0 & B \end{pmatrix}.$$

L is the Jacobi matrix of Φ at $x = 0$; hence L is nonsingular. Therefore there exists $\delta > 0$ such that $|Lx| \geq \delta|x|$ for any x. Find $\varepsilon > 0$ such that

(1) $\varepsilon < \delta$;

(2) $C = \max(\|A\|, \|B^{-1}\|) + \varepsilon + \varepsilon\|B^{-1}\| < 1.$ (A.6)

By Lemma 4.5 there exists a vector function $\tilde{f}$ of class C^1 on $\mathbf{R}^n$ such that

(1) $\tilde{f} = f$ in a neighborhood of the origin;

(2) $\|\frac{\partial \tilde{f}}{\partial x}(x)\| < \varepsilon$ for $x \in \mathbf{R}^n$.

Consider the map $\widetilde{\Phi} : \mathbf{R}^n \to \mathbf{R}^n$ defined by

$$\widetilde{\Phi}(x) = Lx + \tilde{f}(x).$$

We claim that there exists a homeomorphism $h : \mathbf{R}^n \to \mathbf{R}^n$, $h(0) = 0$, conjugating $\widetilde{\Phi}$ and Ψ on $\mathbf{R}^n$. Then h is a local conjugacy between Φ and Ψ. Indeed, let $\widetilde{\Phi} \circ h = h \circ \Psi$. Consider a neighborhoood $\widetilde{U}$ such that $\widetilde{\Phi}$ coincides with Φ in $\widetilde{U}$. Then $\widetilde{U} \cap h^{-1}(\widetilde{U})$ is also a neighborhood of the origin and we have

$$\Phi(h(x)) = \widetilde{\Phi}(h(x)) = h(\Psi(x))$$

for $x \in \widetilde{U} \cap h^{-1}(\widetilde{U})$. To simplify notation we denote $\varphi = \widetilde{\Phi}$ and write $f(x)$ instead of $\tilde{f}(x)$. Let us note that φ is one-to-one. Indeed, if $\varphi(x_1) = \varphi(x_2)$ then

$$Lx_1 + f(x_1) = Lx_2 + f(x_2) \text{ or } L(x_1 - x_2) = f(x_2) - f(x_1).$$

By the choice of δ we have $|L(x_1 - x_2)| \geq \delta|x_1 - x_2|$. As

$$|f(x_2) - f(x_1)| \leq \max_{x \in \mathbf{R}^n} \|\frac{\partial f}{\partial x}(x)\| \, |x_1 - x_2| \leq \varepsilon|x_1 - x_2|$$

we obtain that $\delta|x_1 - x_2| \leq \varepsilon|x_1 - x_2|$. Now (A.6) implies $x_1 = x_2$.

We are going to find a homeomorphism h conjugating φ and Ψ in the form

$$h(x) = x + g(x) \qquad (A.8)$$

where g is a continuous map $\mathbf{R}^n \to \mathbf{R}^n$ having the following additional property:

$$|g(x)| \underset{|x| \to \infty}{\longrightarrow} 0. \qquad (A.9)$$

We denote by I the identity map $\mathbb{R}^n \to \mathbb{R}^n$. If we write (A.8) as $h = I + g$, h is a conjugacy between φ and Ψ if and only if

$$\varphi \circ h = h \circ \Psi \tag{A.10}$$

or

$$\varphi \circ (I + g) = (I + g) \circ \Psi. \tag{A.11}$$

To solve equation (A.11) with respect to g we consider the following space:

$$H = \{g : \mathbb{R}^n \to \mathbb{R}^n : g \in C,\ g(0) = 0, |g(x)| \underset{|x| \to \infty}{\longrightarrow} 0\}.$$

According to the decomposition $\mathbb{R}^n = \mathbb{R}^{n_1} \times \mathbb{R}^{n_2}$ we write $g \in H$ as $g = (g_1, g_2)$. Denote for $g, k \in H$

$$\rho(g, k) = \sup_{x \in \mathbb{R}^n} |g_1(x) - k_1(x)| + \sup_{x \in \mathbb{R}^n} |g_2(x) - k_2(x)|.$$

It is easy to see that ρ is a metric on H, and (H, ρ) is a complete metric space.

Substitute $\Psi(x) = L(x)$, $\varphi(x) = Lx + f(x)$ in (A.11); we obtain

$$(L + f) \circ (I + g) = (I + g) \circ L$$

or

$$L + L \circ g + f \circ (I + g) = L + g \circ L. \tag{A.12}$$

Note that as L is linear, we have $L \circ (I + g) = L + L \circ g$ but $f \circ (I + g) \neq f \circ I + f \circ g$! (A.12) is equivalent to

$$L \circ g + f \circ (I + g) = g \circ L. \tag{A.13}$$

Let us write (A.13) at $x \in \mathbb{R}^n$:

$$Lg(x) + f(x + g(x)) = g(Lx).$$

The operator L is invertible; hence (A.13) is equivalent to

$$g = L^{-1} \circ (g \circ L - f \circ (I + g)). \tag{A.14}$$

According to the decomposition $x = (y, z)$ and taking into account that

$$L^{-1} = \begin{pmatrix} A^{-1} & 0 \\ 0 & B^{-1} \end{pmatrix}$$

we can write (A.14) as

$$g_1 = A^{-1} \circ [g_1 \circ L - f_1 \circ (I+g)], \tag{A.15}$$

$$g_2 = B^{-1} \circ [g_2 \circ L - f_2 \circ (I+g)]. \tag{A.16}$$

Multiplying (A.15) by A and taking $L^{-1}x$ instead of x we obtain

$$g_1 = A \circ g_1 \circ L^{-1} + f_1 \circ (I+g) \circ L^{-1}.$$

Define the operator T: for $g \in H$

$$T(g) = (A \circ g_1 \circ L^{-1} + f_1 \circ (I+g) \circ L^{-1}, B^{-1} \circ [g_2 \circ L - f_2 \circ (I+g)]).$$

Evidently $T(g) \in C(\mathbb{R}^n)$ and $T(g)(0) = 0$. As $|g(x)| \to 0$ for $|x| \to \infty$ we have $|(I+g)(x)| = |x + g(x)| \to \infty$ for $|x| \to \infty$. It follows from the construction of f that $f(x) = 0$ for large $|x|$ (see Lemma 4.5), and hence

$$|f_2 \circ (I+g)(x)| \to 0 \text{ for } |x| \to \infty.$$

If $|x| \to \infty$ then $|Lx| \to \infty$; therefore $|g_2(Lx)| \to 0$ and $|B^{-1}g_2(Lx)| \to 0$ for $|x| \to \infty$. So the norm of the second component of $T(g)$ tends to zero as $|x| \to \infty$. Similar considerations show that the first component of $T(g)$ has analogous properties. So $T(g) \in H$, and hence T is a map $H \to H$. Let us show that T is a contraction on H. For $g, k \in H$ we have

$$\begin{aligned}\rho(T(g), T(k)) &= \sup_{x \in \mathbb{R}^n} \{|Ag_1(L^{-1}x) - Ak_1(L^{-1}x) \\ &+ f_1(L^{-1}x + g(L^{-1}x)) - f_1(L^{-1}x + k(L^{-1}x))|\} \\ &+ \sup_{x \in \mathbb{R}^n} \{|B^{-1}[g_2(Lx) - k_2(Lx) \\ &+ f_2(x + k(x)) - f_2(x + g(x))]|\}.\end{aligned} \tag{A.17}$$

Note that as L is invertible,

$$\sup_{x \in \mathbb{R}^n} |A(g_1(L^{-1}x) - k_1(L^{-1}x))| = \sup_{x \in \mathbb{R}^n} |A(g_1(x) - k_1(x))|.$$

Therefore we can estimate the first term in (A.17) from above by

$$\begin{aligned}&\|A\| \sup_{x \in \mathbb{R}^n} |g_1(x) - k_1(x)| + \sup_{x \in \mathbb{R}^n} \|\frac{\partial f_1}{\partial x}\| |g(x) - k(x)| \\ &\le \|A\| \sup_{x \in \mathbb{R}^n} |g_1(x) - k_1(x)| + \varepsilon \sup_{x \in \mathbb{R}^n} |g(x) - k(x)|.\end{aligned}$$

We take into account (A.7) here. Similarly the second term in (A.17) is bounded from above by

$$\|B^{-1}\| \sup_{x\in\mathbb{R}^n} |g_2(x) - k_2(x)| + \varepsilon\|B^{-1}\| \sup_{x\in\mathbb{R}^n} |g(x) - k(x)|.$$

Evidently

$$|g(x) - k(x)| \leq |g_1(x) - k_1(x)| + |g_2(x) - k_2(x)|,$$

hence

$$\rho(T(g), T(k)) \leq [\max(\|A\|, \|B^{-1}\|) + \varepsilon(1 + \|B^{-1}\|)]\rho(g, k).$$

It follows now from (A.6) that T is a contraction. Hence T has a unique fixed point g^* in H. We obtain from the definition of T that $g^* = (g_1^*, g_2^*)$ is a solution of systems (A.15) and (A.16); therefore g^* satisfies (A.11). So, the map $h(x) = x + g^*(x)$ satisfies (A.10). It remains now to prove that h is a homeomorphism. Evidently $h \in C(\mathbb{R}^n)$.

Let us show that h is invertible. To get a contradiction suppose that there exists $x_1, x_2 \in \mathbb{R}^n$ such that $x_1 \neq x_2$, $h(x_1) = h(x_2)$. It was shown in Section 2, Chapter 2 that (A.10) implies

$$\varphi^k \circ h = h \circ \Psi^k, \; k \in \mathbf{Z}.$$

As it follows from our assumption that

$$\varphi^k(h(x_1)) = \varphi^k(h(x_2))$$

we obtain that

$$h(\Psi^k(x_1)) = h(\Psi^k(x_2)),$$

and hence

$$h(L^k x_1) = h(L^k x_2), \; k \in \mathbf{Z}. \tag{A.18}$$

As $h(x) = x + g^*(x)$ we obtain from (A.18) that for $k \in \mathbf{Z}$ we have

$$L^k x_1 + g^*(L^k x_1) = L^k x_2 + g^*(L^k x_2)$$

or

$$L^k(x_1 - x_2) = g^*(L^k x_2) - g^*(L^k x_1). \tag{A.19}$$

Let $x_1 - x_2 = v = (v_1, v_2)$, $v_1 \in \mathbf{R}^{n_1}$, $v_2 \in \mathbf{R}^{n_2}$. By our assumption $v \neq 0$, let $v_1 \neq 0$. Let us show that

$$|A^k v_1| \underset{k \to -\infty}{\longrightarrow} \infty. \tag{A.20}$$

If (A.20) is not true then for the sequence $|A^k v_1|$ has a bounded subsequence. Suppose for simplicity that $|A^k v_1| \leq m$. As $\|A\| < 1$ we have

$$|v_1| = |A^{-k} A^k v_1| \leq \|A\|^{-k} m \underset{k \to -\infty}{\longrightarrow} 0$$

which contradicts the condition $v_1 \neq 0$. So, (A.20) is established. It is easy to see that $|L^k v| \geq |A^k v_1|$, so if $v_1 \neq 0$ then the left part of (A.19) is unbounded. It follows from the definition of H that the right part of (A.19) is bounded. The contradiction we obtained proves that h is invertible.

As h is invertible and continuous, we obtain that h is a homeomorphism on any compact subset of $\mathbf{R}^n$. By the Brower Theorem [9] h maps open sets onto open sets, hence $h(\mathbf{R}^n)$ is open. Let us show that $h(\mathbf{R}^n)$ is closed.

Consider a sequence $h(x_k)$ such that $h(x_k) \underset{k \to \infty}{\longrightarrow} x_0$. The sequence x_k is bounded. Indeed if x_k has an unbounded subsequence x_{k_s} then $g^*(x_{k_s}) \to 0$ and

$$|h(x_{k_s})| = |x_{k_s} + g^*(x_{k_s})| \to \infty.$$

Find a convergent subsequence x_{k_m} of x_{k_m}. Let $x_{k_m} \to \tilde{x}$. It follows from the continuity of h that$h(x_{k_m}) \to h(\tilde{x})$; therefore $x_0 = h(\tilde{x}) \in h(\mathbf{R}^n)$. We proved that $h(\mathbf{R}^n)$ is closed so we see that $h(\mathbf{R}^n) = \mathbf{R}^n$. That completes the proof of our theorem. □

To prove Theorem 4.3 denote by $\Psi(t, x)$ the flow generated by system (4.47). It follows from the proof of the λ-Lemma for flows (see Chapter 6) that there exists $\theta > 0$ such that the diffeomorphism $\varphi(\theta, x)$ (here φ is the flow of (1.1)) satisfies the conditions of the theorem we proved. Hence, the fixed point $x = 0$ of $\varphi(\theta, x)$ is locally topologically conjugate to the fixed point $x = 0$ of the diffeomorphism $\Psi(\theta, x)$. Denote by h_0 the corresponding homeomorphism; then there exists a neighborhood U_0 of the origin such that

$$h_0(\varphi(\theta, x)) = \Psi(\theta, h_0(x)) \tag{A.21}$$

for $x \in U_0$. Denote by $\varphi(t,\cdot), \Psi(t,\cdot)$ the restrictions of flows φ, Ψ on $\{t\} \times \mathbf{R}^n$ (i.e. $\varphi(t,\cdot)(x) = \varphi(t,x)$) and define the following map:

$$h = \int_0^\theta \Psi(-s,\cdot) \circ h_0 \circ \varphi(s,\cdot) ds.$$

Note that

$$h(x) = \int_0^\theta \Psi(-s, h_0(\varphi(s,x))) ds.$$

Let us show that (2.5) holds for t, x such that $\varphi(t,x) \in U_0$.

Fix t and consider

$$\Psi(t,\cdot) \circ h = \Psi(t,\cdot) \circ \int_0^\theta \Psi(-s,\cdot) \circ [\ldots] ds =$$

(we write $[\ldots]$ for $h_0 \circ \varphi(s,\cdot)$)

$$\begin{aligned}
&= e^{At} \int_0^\theta e^{-As} \circ [\ldots] ds = \int_0^\theta e^{A(t-s)} \circ [\ldots] ds \\
&= \int_0^\theta \Psi(t-s,\cdot) \circ h_0 \circ \varphi(s,\cdot) ds \\
&= (\int_0^\theta \Psi(t-s,\cdot) \circ h_0 \circ \varphi(s-t,\cdot) ds) \circ \varphi(t,\cdot). \qquad (A.22)
\end{aligned}$$

We take into account that $\varphi(s,\cdot) = \varphi(s-t,\cdot) \circ \varphi(t,\cdot)$.

Consider the integral in brackets in (A.22). By putting $s - t = \sigma$ we obtain

$$\begin{aligned}
\int_0^\theta (\quad) ds &= \int_{-t}^{\theta - t} \Psi(-\sigma,\cdot) \circ h_0 \circ \varphi(\sigma,\cdot) d\sigma \\
&= \int_{-t}^0 (\quad) d\sigma + \int_0^{\theta - t} (\quad) d\sigma. \qquad (A.23)
\end{aligned}$$

Let us transform

$$\begin{aligned}
&\Psi(-\sigma,\cdot) \circ h_0 \circ \varphi(\sigma,\cdot) \\
&= \Psi(-\sigma - \theta,\cdot) \circ \Psi(\theta,\cdot) \circ h_0 \circ \varphi(\sigma,\cdot) \\
&= \Psi(-\sigma - \theta,\cdot) \circ h_0 \circ \varphi(\theta,\cdot) \circ \varphi(\sigma,\cdot) \qquad (A.24)
\end{aligned}$$

(we use (A.21) in the following form:

$$\begin{aligned}\Psi(\theta,\cdot)\circ h_0 &= h_0\circ\varphi(\theta,\cdot)) \\ &= \Psi(-\sigma-\theta,\cdot)\circ h_0\circ\varphi(\theta+\sigma,\cdot)\end{aligned} \tag{A.25}$$

Put $\tau=\theta+\sigma$ in the first integral in (A.23) and note that (A.24) equals (A.25):

$$\begin{aligned}&\int_{-t}^{0}\Psi(-\sigma,\cdot)\circ h_0\circ\varphi(\sigma,\cdot)d\sigma \\ &= \int_{-t}^{0}\Psi(-\sigma-\theta,\cdot)\circ h_0\circ\varphi(\theta+\sigma,\cdot)d\sigma \\ &= \int_{\theta-t}^{\theta}\Psi(-\tau,\cdot)\cdot h_0\circ\varphi(\tau,\cdot)d\tau.\end{aligned} \tag{A.26}$$

We see that the integrands of (A.23) and (A.26) differ only in notation; hence

$$\begin{aligned}&\int_{0}^{\theta}\Psi(t-s,\cdot)\circ h_0\circ\varphi(s-t,\cdot)ds \\ &= \int_{0}^{\theta}\Psi(-s,\cdot)\circ h_0\circ\varphi(s,\cdot)ds = h.\end{aligned}$$

Finally, we see that $\Psi(t,\cdot)\circ h = h\circ\varphi(t,\cdot)$. To show that h is a homeomorphism the reader can repeat the end of the proof of the theorem preceding the proof of the Grobman–Hartman Theorem.

References

1. A.A. Andronov, L.S. Pontryagin, *Systèmes grossiers*, C.R. (Dokl.) Acad. Sci. USSR. **14**(1937), 247–251.
2. D.V. Anosov, *Geodesic Flows on Closed Riemannian Manifolds of Negative Curvature*, Proc. Steklov Inst. Math. **90** (1967).
3. V.I. Arnold, *Ordinary Differential Equations* (in Russian); В. И. Арнольд, Обыкновенные Дифференциальные уравнения. Наука, М., 1975.
4. J. Franks, *Homology and Dynamical Systems*, Reg. Conf. Ser. in Math., Conf. Board of Math.Sci., AMS, 1982.
5. D.M. Grobman, *Topological Classification of Neighborhoods of a Singular Point in n-dimensional Space* (in Russian); Д.М. Гробман. Топологическая классификация окрестностей особои точки в n-мерном пространстве. Матем. сборник, **56** 1, (1962), 77-94.
6. P. Hartman, *Ordinary Differential Equations*, New York: Wiley, 1964.
7. M. Hirsch, J. Palis, C. Pugh, M. Shub, *Neighborhoods of Hyperbolic Sets*, Invent. math. **2** (1970), 121–134 **14**(1970), 133–163.
8. M. Hirsch, C. Pugh, *Stable Manifolds and Hyperbolic Sets*, Glob. Analysis Symp. in Pure Math., **14** (1970), 133–163.
9. W. Hurewicz, H. Wallman, *Dimension Theory*, 1941 (in Russian); Б. Гуревич, Г. Волман. Теория размерности. Москва, 1948.
10. M.A. Krasnoselsky, A.I. Perov, A.I. Povolotsky, P.E. Zabreyko, *Vector fields in the plane*, (in Russian); М.А. Красносельский, А.И. Перов, А.И. Поволоцкий, П.Е. Забрейко, Векторные поля на плоскости, М., 1963.
11. I. Kupka, *Contributions à la théorie des champs génériques*, Contrib. to Diff. Equat. **2**(1963), 457–484.
12. R. Mañé, *Characterizations of AS Diffeomorphisms*, Geom. and Top. III, Lat. Am. Sch. of Math., July 1976, Lect. Notes in Math., **597** (1977), 389–394.
13. R. Mañé, *A Proof of the C^1 Stability Conjecture*, IHES Pub. Math. **66**(1988), 161–210.
14. E. Markus, *Lectures in Differentiable Dynamics*, Reg. Conf. Ser. in Math., Conf. Board of the Math. Sci., AMS, 1971.
15. R. Narasimhan, *Analysis on Real and Complex Manifolds*, Paris: Masson & Cie, Paris, 1968.
16. Yu.I. Neimark, *On movements close to double-asymptotic movement*, (in Russian); Ю.И. Неймарк. О движениях, близких к двояко-асимптотическому движению. Доклады АН СССР, **172** 5, (1967), 1021–1024.

17. V.V. Nemyckii, V.V. Stepanov, *Qualitative theory of differential equations*, Princeton Math. Series **22** (1960), Princeton, N.J.
18. Z. Nitecki, *Differentiable Dynamics*, Cambridge, MA: MIT Press, 1971.
19. J. Palis, *On Morse-Smale Dynamical Systems*, Topology, **8** 4, (1969), 385–404.
20. J. Palis, S. Smale, *Structural Stability Theorems*, Glob. Analysis Symp. in Pure Math., **14** (1977), 223-231.
21. J. Palis, *Vector Fields Generate Few Diffeomorphisms*, Bull. Amer. Math. Soc. **803**, (1974), 503–505.
22. J. Palis, W. de Melo, *Geometric Theory of Dynamical Systems*, Springer-Verlag, 1982.
23. J. Palis, *On the C^1 Ω-stability conjecture*, IHES Pub. Math. **66** (1988), 211–215.
24. V.A. Pliss, *Integral sets of periodic systems of differential equations* (in Russian); В.А. Плисс. Интеграьные множества периодических систем дифференциальних уравнений. М., 1977.
25. Ch. Pugh, *The Closing Lemma*, Amer. J. Math., **89**(1967), 956–1009.
26. Ch. Pugh, *An Improved Closing Lemma and a General Density Theorem*, Am. Jour. Math., **89** (1967), 1010–1021.
27. L. Reizins, *Local equivalence of differenential equations* (in Russian); Л. Реизинь. Локальная эквивалентностьдифференциальных уравнений. Рига, 1971.
28. J. Robbin, *A structural stability theorem*, Ann. Math. **94** 3, (1971), 447–493.
29. C. Robinson, *Structural stability of C^1-flows, Dynamical Systems,* Warwick 1974, Lect. Notes Math., **468** (1975), Springer- Verlag.
30. C. Robinson, *Structural stability for C^1-diffeomorphisms*, Journ. Diff. Eq., **22** 1, (1976), 28–73.
31. C. Robinson, *Introduction to the Closing Lemma: The Structure of Attractors in Dynamical Systems*, Lect. Notes Math., **668** (1978), Springer-Verlag.
32. V.A. Rohlin, D.B. Fuks, *First course of Topology. Geometric chapters* (in Russian); В.А. Рохлинь, Д.Б. Фукс. Начальныи курс топологии. Геометрические главы. М., 1977.
33. R.J. Sacker, G.R. Sell, *Existence of dichotomies and invariant splittings for linear differential systems. I-II.* J. Diff. Eq., **15** 3, (1974), 492–458; **22** 2, (1976), 478–496.
34. L.P. Shilnikov, *On a problem of Poincaré-Birkhoff*, (in Russian); Л.П. Шилников. Об однои задаче Пуанкаре-Биркгофа. Матем сборник **74** 3, (1967), 378–397.
35. M. Shub, *Stabilité globale des systèmes dynamiques*, Astérisque, **56** (1978), 1–211.
36. S. Smale, *Morse Inequalities for a Dynamical System*, Bull. Amer. Math. Soc., **66** (1960), 43–49.

37. S. Smale, *Stable Manifolds for Differential Equations and Diffeomorphisms*, Ann. Schuola Norm. Sup. Pisa **17** 3, (1963), 97–116.
38. S. Smale, *Diffeomorphisms with Many Periodic Points*, Differential and Combinatorial Topology. Princeton Univ. Press, Princeton, NJ 1965, 63–80.
39. S. Smale, *Differentiable Dynamical Systems*, Bull. Amer. Math. Soc., **73** (1967), 747–817.
40. S. Smale, *The Ω-Stability Theorem*, Glob. Analysis Symp. in Pure Math., **14** (1970), 289–297.

Index